Recent volumes in MSJ Memoirs

1 **Ivan Cherednik, Peter J. Forrester and Denis Uglov**, Quantum many-body problems and representation theory, 1998
2 **Masako Takahashi, Mitsuhiro Okada and Mariangiola Dezani-Ciancaglini (Eds.)**, Theories of types and proofs, 1998
3 **Tomotada Ohtsuki**, Combinatorial quantum method in 3-dimensional topology, 1999
4 **Atsushi Matsuo and Kiyokazu Nagatomo**, Axioms for a vertex algebra and the locality of quantum fields, 1999
5 **Daryl Cooper, Craig D. Hodgson and Steven P. Kerckhoff**, Three-dimensional orbifolds and cone-manifolds, 2000
6 **Kong De-xing**, Cauchy problem for quasilinear hyperbolic systems, 2000
7 **Vladimir Georgiev**, Semilinear hyperbolic equations, 2000
8 **Eric M. Opdam**, Lecture notes on Dunkl operators for real and complex reflection groups, 2000
9 **Peter Orlik and Hiroaki Terao**, Arrangements and hyper-geometric integrals, 2001
10 **Yuri G. Prokhorov**, Lectures on complements on log surfaces, 2001
11 **John R. Stembridge, Jean-Yves Thibon and Marc A. A. van Leeuwen**, Interaction of combinatorics and representation theory, 2001
12 **Yves André**, Period mappings and differential equations. From $\mathbb{C}$ to $\mathbb{C}_p$, 2003
13 **Shigeaki Koike**, A beginner's guide to the theory of viscosity solutions, 2004
14 **Noboru Nakayama**, Zariski-decomposition and Abundance, 2004
15 **Valery Alexeev and Viacheslav V. Nikulin**, Del Pezzo and K3 surfaces, 2006
16 **Brian H. Bowditch**, A course on geometric group theory, 2006
17 **Arkady Berenstein, David Kazhdan, Cédric Lecouvey, Masato Okado, Anne Schilling, Taichiro Takagi and Alexander Veselov**, Combinatorial Aspect of Integrable Systems, 2007
18 **Yasutaka Ihara**, On Congruence Monodromy Problems, 2008
19 **Joseph Najnudel, Bernard Roynette and Marc Yor**, A Global View of Brownian Penalisations, 2009
20 **Danny Calegari**, scl (Stable Commutator Length), 2009
21 **Gautami Bhowmik, Kohji Matsumoto and Hirofumi Tsumura (Eds.)**, Algebraic and Analytic Aspects of Zeta Functions and L-functions, 2010
22 **Michael Ruzhansky and James Smith**, Dispersive and Strichartz estimates for hyperbolic equations with constant coefficients, 2010
23 **Armen Sergeev**, Kähler geometry of loop spaces, 2010
24 **Taro Asuke**, Godbillon–Vey class of transversely holomorphic foliations, 2010
25 **Hiroshi Sugita**, Monte Carlo method, random number, and pseudorandom number, 2011
26 **Shinya Nishibata and Masahiro Suzuki**, Hierarchy of semiconductor equations: relaxation limits with initial layers for large initial data, 2011
27 **Suhyoung Choi**, Geometric structures on 2-orbifolds: Exploration of discrete symmetry, 2012

28 **Toshio Oshima**, Fractional calculus of Weyl algebra and Fuchsian differential equations, 2012

29 **Takeshi Hirai, Akihito Hora and Etsuko Hirai**, Projective representations and spin characters of complex reflection groups $G(m, p, n)$ and $G(m, p, \infty)$, 2013

30 **Tatsuo Nishitani**, Cauchy Problem for Noneffectively Hyperbolic Operators, 2013

31 **Satoshi Takanobu**, Bohr–Jessen Limit Theorem, Revisited, 2013

32 **Hiroshi Isozaki and Yaroslav Kurylev**, Introduction to spectral theory and inverse problem on asymptotically hyperbolic manifolds, 2014

33 **Masaki Maruyama**, Moduli spaces of stable sheaves on schemes, 2016

34 **Martin T. Barlow, Tibor Jordán and Andrzej Zuk**, Discrete Geometric Analysis, 2016

35 **Osamu Fujino**, Foundations of the minimal model program, 2017

36 **Soichiro Katayama**, Global solutions and the asymptotic behavior for nonlinear wave equations with small initial data, 2018

37 **Kazuki Hiroe, Hiroshi Kawakami, Akane Nakamura and Hidetaka Sakai**, 4-dimensional Painlevé-type equations, 2018

38 **Alex Casella, Dominic Tate and Stephan Tillmann**, Moduli spaces of real projective structures on surfaces, 2020

39 **Masaharu Taniguchi**, Traveling Front Solutions in Reaction-Diffusion Equations, 2021

40 **Yuji Odaka and Yoshiki Oshima**, Collapsing K3 Surfaces, Tropical Geometry and Moduli Compactifications of Satake, Morgan–Shalen Type, 2021

41 **Tomoki Nakanishi**, Cluster Algebras and Scattering Diagrams, 2023

42 **Osamu Fujino**, Cone and contraction theorem for projective morphisms between complex analytic spaces, 2024

Cone and contraction theorem for projective morphisms between complex analytic spaces

Osamu Fujino

MSJ Memoirs

Mathematical Society of Japan

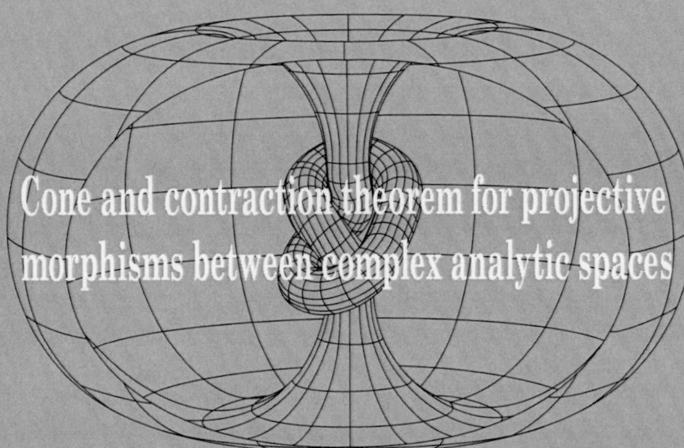

Cone and contraction theorem for projective morphisms between complex analytic spaces

Osamu Fujino

Volume 42 2024

Author
Osamu Fujino
Department of Mathematics, Graduate School of Science, Kyoto University, Kyoto 606-8502, Japan
fujino@math.kyoto-u.ac.jp

Keywords: minimal model program, log canonical pairs, complex analytic spaces, cone and contraction theorem, basepoint-free theorem, extremal rational curves, vanishing theorems, strict support condition

2020 Mathematics Subject Classifications: Primary 14E30; Secondary 32C15.

MSJ Memoirs

This monograph series is intended to publish lecture notes, graduate textbooks and long research papers* in pure and applied mathematics. Each volume should be an integrated monograph. Proceedings of conferences or collections of independent papers are not accepted.

Articles for the series can be submitted to one of the editors in the form of hard copy. When the article is accepted, the author(s) is (are) requested to send a camera-ready manuscript.

* long research papers are limited to contributions by MSJ members

Editorial Board

The series *MSJ Memoirs* is published occasionally (3–5 volumes each year) by the Mathematical Society of Japan. The titles and prices of recent volumes are listed on page i of this volume. Each volume of the series can be separately purchased by placing an order with:

- The Mathematical Society of Japan, 1-34-8 Taito, Taito-ku, 110-0016 Tokyo, Japan (from inside Japan)
- World Scientific Publishing, 5 Thoh Tuck Link, Singapore 596224 (from outside Japan)

For Yoko, Kentaro, and Makoto

Preface

We discuss the cone and contraction theorem in a suitable complex analytic setting. More precisely, we establish the cone and contraction theorem of normal pairs for projective morphisms between complex analytic spaces. This result is a starting point of the minimal model program for complex analytic log canonical pairs. In this book, we are mainly interested in normal pairs whose singularities are worse than kawamata log terminal singularities. This book is intended to give a rigorous foundation of the theory of minimal models for projective morphisms of complex analytic spaces.

Acknowledgments. First of all, the author thanks the referee very much for many useful comments and suggestions. He thanks Professor Noboru Nakayama very much for useful suggestions and answering his questions. He thanks Natsuo Saito for drawing a beautiful picture of a Kleiman–Mori cone. He also would like to thank Professors Taro Fujisawa, Shigefumi Mori, and Morihiko Saito very much. Finally, he thanks Kenta Hashizume for many useful comments.

During the preparation of this book, he was partially supported by JSPS KAKENHI Grant Numbers JP19H01787, JP20H00111, JP21H00974, JP21H04994, and JP23K20787.

August 26, 2024
Osamu Fujino

Contents

Chapter 1
Introduction

In Section 1.1, we will explain the main results of this book. In Section 1.2, we will shortly discuss some conjectures on the theory of minimal models for projective morphisms between complex analytic spaces. In Section 1.3, we will explain some examples which show various differences between algebraic varieties and complex analytic spaces. In Section 1.4, we will describe the contents of each chapter more precisely for the reader's convenience.

1.1 Main results

In his epoch-making paper [Mo], Shigefumi Mori established the cone theorem for smooth projective varieties defined over any algebraically closed field k of arbitrary characteristic by his ingenious method of *bend and break*. Then he established the contraction theorem for smooth projective threefolds when the characteristic of the base field k is zero. After that, in characteristic zero, the cone and contraction theorem was generalized for so-called log-terminal pairs in any dimension by using Hironaka's resolution of singularities and the Kawamata–Viehweg vanishing theorem. For the details, see [KMM], [KM], [Ma] and references therein. Now we know that, in characteristic zero, the cone and contraction theorem holds for more general settings (see [Fu4], [Fu6, Chapter 6], [Fu11], and references therein). In this book, we will discuss the cone and contraction theorem of normal pairs for projective morphisms between complex analytic spaces. For kawamata log terminal pairs, it was known and has played an important role in [Na1], [Na2], and [Fu13]. In [Fu13], we have already discussed the minimal model program for kawamata log terminal pairs in a complex analytic setting (see also [Fu17]). Roughly speaking, we showed that [BCHM] and [HM] can work for projective morphisms between complex analytic spaces. We note that the Kawamata–Viehweg vanishing theorem can be formulated and proved for projective morphisms of complex analytic spaces and is sufficient for the study of kawamata log terminal pairs. We also note that L^2-methods can work for kawamata log terminal pairs. For an alternative approach to the minimal model program of kawamata log terminal pairs for projective morphisms between complex analytic spaces, see [DHP], which uses the idea of [CL]. The reader can find a new approach to the relative minimal model program in [LM], which can work in larger categories of spaces. In [Fu15], we established some vanishing theorems and related results necessary for the study of complex analytic log canonical pairs and quasi-log structures on complex analytic spaces.

Theorem 1.1.1 ([Fu15, Theorem 1.1], see also Theorem 3.1.5)**.** *Let (X, Δ) be an analytic simple normal crossing pair such that Δ is a boundary $\mathbb{R}$-divisor on X. Let*

$f\colon X \to Y$ be a projective morphism to a complex analytic space Y and let $\mathcal{L}$ be a line bundle on X. Let q be an arbitrary non-negative integer. Then we have the following properties.

(1) (*Strict support condition*). *If $\mathcal{L} - (\omega_X + \Delta)$ is f-semiample, then every associated subvariety of $R^q f_* \mathcal{L}$ is the f-image of some stratum of (X, Δ).*

(2) (*Vanishing theorem*). *If $\mathcal{L} - (\omega_X + \Delta) \sim_{\mathbb{R}} f^* \mathcal{H}$ holds for some π-ample $\mathbb{R}$-line bundle $\mathcal{H}$ on Y, where $\pi\colon Y \to Z$ is a projective morphism to a complex analytic space Z, then we have*

$$R^p \pi_* R^q f_* \mathcal{L} = 0$$

for every $p > 0$.

Theorem 1.1.1 in the usual algebraic setting is well known and has already played an important role in the theory of minimal models (see, for example, [Fu6, Chapters 5 and 6]). Hence it is very natural to consider Theorem 1.1.1 for the minimal model theory of projective morphisms between complex analytic spaces. The approach to Theorem 1.1.1 adopted in [Fu6, Chapter 5] does not work for complex analytic spaces. Therefore, the proof of Theorem 1.1.1 in [Fu15] is completely different from the corresponding one in the algebraic setting in [Fu6, Chapter 5]. Note that [Fu15] depends on Morihiko Saito's theory of mixed Hodge modules (see [Sa1], [Sa2], [Sa3], [FFS], and [Sa4]) and Takegoshi's generalization of Kollár's torsion-free and vanishing theorem (see [Ta]). For an approach to Theorem 1.1.1 without using the theory of mixed Hodge modules, see [FF], where we use the theory of variations of mixed Hodge structure. As in the algebraic case (see [Fu4] and [Fu6]), our philosophy is as follows.

1.1.2 (Philosophy). From the Hodge theoretic viewpoint, we have the following correspondences.

Kawamata log terminal pairs $\Longleftrightarrow$ Pure Hodge structures

and

Quasi-log pairs $\Longleftrightarrow$ Mixed Hodge structures

For a completely different approach to Theorem 1.1.1 due to Murayama, see [Mu]. We note that his new approach can work for schemes, algebraic stacks, formal schemes, semianalytic germs of complex analytic spaces, and so on.

In this book, we will discuss the cone and contraction theorem of normal pairs for projective morphisms between complex analytic spaces as an application of Theorem 1.1.1. More generally, in [Fu16], we will discuss quasi-log structures for complex analytic spaces. We can see this book as a complex analytic generalization of [Fu4] and as a generalization of Nakayama's paper [Na1]. We note that Nakayama only treated kawamata log terminal pairs and $\mathbb{Q}$-divisors in [Na1]. Finally, this book is independent of [Fu13] and does not use any results obtained in [Fu13].

This book is intended to give a rigorous foundation of the theory of minimal models for projective morphisms of complex analytic spaces.

1.1.3 (Standard setting)**.** One of the main difficulties to discuss the minimal model theory for complex analytic spaces is how to formulate it.

Let $\pi\colon X \to Y$ be a projective morphism between complex analytic spaces such that X is a normal complex variety and let W be a compact subset of Y. In this book, we formulate and prove almost everything over some open neighborhood of W. Let Δ be an $\mathbb{R}$-divisor on X such that $K_X+\Delta$ is $\mathbb{R}$-Cartier. The number of the irreducible components of $\operatorname{Supp}\Delta$ is only locally finite. In general, the support of Δ may have infinitely many irreducible components. By shrinking Y around W suitably, that is, by replacing Y with a suitable relatively compact open neighborhood of W, we can always assume that $\operatorname{Supp}\Delta$ has only finitely many irreducible components. Moreover, we can assume that a given $\mathbb{R}$-Cartier divisor on X is a finite $\mathbb{R}$-linear combination of Cartier divisors. Therefore, by considering some relatively compact open neighborhood of W, we can avoid subtle problems caused by the difference between the Zariski topology and the Euclidean topology. In [Fu13], we almost always assume that W is a Stein compact subset of Y such that $\Gamma(W,\mathcal{O}_Y)$ is noetherian. In this book, however, we usually assume that W is only a compact subset of Y. When we consider the Kleiman–Mori cone $\overline{\mathrm{NE}}(X/Y;W)$ of $\pi\colon X\to Y$ and W, we further assume that the dimension of $N^1(X/Y;W)$ is finite. For the details of $\overline{\mathrm{NE}}(X/Y;W)$ and $N^1(X/Y;W)$, see Section 4.4. Note that if $W\cap V$ has only finitely many connected components for any analytic subset V which is defined over an open neighborhood of W then the dimension of $N^1(X/Y;W)$ is finite by Nakayama's finiteness (see Theorem 4.5.2). Therefore, if W is a compact semianalytic subset of Y, then the dimension of $N^1(X/Y;W)$ is always finite. Thus, we can find many compact subsets W with $\dim_{\mathbb{R}} N^1(X/Y;W)<\infty$.

We make an important remark.

Remark 1.1.4. Although some statements in this book may hold globally, we basically formulate and prove them only on relatively compact open subsets. This is mainly because it is sufficient for our purposes and it is psychologically easier. We do not like to treat infinite sequences of blow-ups, divisors which have infinitely many irreducible components, and so on.

Although we are mainly interested in normal complex varieties in this book, we frequently have to treat disjoint unions of normal complex varieties.

Remark 1.1.5. Let $\pi\colon X\to Y$ be a projective morphism from a normal complex variety X to a complex analytic space Y and let W be a compact subset of Y as in 1.1.3. Throughout this book, we frequently shrink Y around W. More precisely, we replace Y with an open neighborhood U of W. If the natural map $\mathcal{O}_Y\to\pi_*\mathcal{O}_X$ is an isomorphism, then $\pi^{-1}(U)$ is connected for every connected open subset U of Y. However, if general fibers of $\pi\colon X\to Y$ are not connected, then $\pi^{-1}(U)$ may become a disjoint union of normal complex varieties even when U is connected. Moreover, if π is surjective and U is not connected, then $\pi^{-1}(U)$ is obviously a disjoint union of normal complex varieties.

Main theorem: In this book, we call (X,Δ) a *normal pair* if it consists of a normal complex variety X and an effective $\mathbb{R}$-divisor Δ on X such that $K_X+\Delta$ is $\mathbb{R}$-Cartier. The main purpose of this book is to establish the following cone and contraction theorem of normal pairs for projective morphisms between complex analytic spaces.

Theorem 1.1.6 (Main theorem, cone and contraction theorem, see Theorems 4.6.1, 4.6.2, 5.1.3, and 5.2.4). *Let $\pi\colon X \to Y$ be a projective morphism of complex analytic spaces such that X is a normal complex variety and let W be a compact subset of Y. Assume that the dimension of $N^1(X/Y;W)$ is finite. Let Δ be an effective $\mathbb{R}$-divisor on X such that $K_X+\Delta$ is $\mathbb{R}$-Cartier. Then we have*

$$\overline{\mathrm{NE}}(X/Y;W)=\overline{\mathrm{NE}}(X/Y;W)_{(K_X+\Delta)\geq 0}+\overline{\mathrm{NE}}(X/Y;W)_{\mathrm{Nlc}(X,\Delta)}+\sum R_j$$

with the following properties.

(1) $\mathrm{Nlc}(X,\Delta)$ *is the non-lc locus of* (X,Δ) *and* $\overline{\mathrm{NE}}(X/Y;W)_{\mathrm{Nlc}(X,\Delta)}$ *is the subcone of* $\overline{\mathrm{NE}}(X/Y;W)$ *which is the closure of the convex cone spanned by the projective integral curves* C *on* $\mathrm{Nlc}(X,\Delta)$ *such that* $\pi(C)$ *is a point of* W.

(2) R_j *is a* $(K_X+\Delta)$*-negative extremal ray of* $\overline{\mathrm{NE}}(X/Y;W)$ *which satisfies*

$$R_j\cap\overline{\mathrm{NE}}(X/Y;W)_{\mathrm{Nlc}(X,\Delta)}=\{0\}$$

for every j.

(3) *Let* $\mathcal{A}$ *be a* π*-ample* $\mathbb{R}$*-line bundle on* X. *Then there are only finitely many* R_j*'s included in* $\overline{\mathrm{NE}}(X/Y;W)_{(K_X+\Delta+\mathcal{A})<0}$. *In particular, the* R_j*'s are discrete in the half-space* $\overline{\mathrm{NE}}(X/Y;W)_{(K_X+\Delta)<0}$.

(4) *Let* F *be any face of* $\overline{\mathrm{NE}}(X/Y;W)$ *such that*

$$F\cap\left(\overline{\mathrm{NE}}(X/Y;W)_{(K_X+\Delta)\geq 0}+\overline{\mathrm{NE}}(X/Y;W)_{\mathrm{Nlc}(X,\Delta)}\right)=\{0\}.$$

Then, after shrinking Y *around* W *suitably, there exists a contraction morphism* $\varphi_F\colon X\to Z$ *over* Y *satisfying the following properties.*

   (i) *Let* C *be a projective integral curve on* X *such that* $\pi(C)$ *is a point of* W. *Then* $\varphi_F(C)$ *is a point if and only if the numerical equivalence class* $[C]$ *of* C *is in* F.

   (ii) *The natural map* $\mathcal{O}_Z\to(\varphi_F)_*\mathcal{O}_X$ *is an isomorphism.*

   (iii) *Let* $\mathcal{L}$ *be a line bundle on* X *such that* $\mathcal{L}\cdot C=0$ *for every curve* C *with* $[C]\in F$. *Then, after shrinking* Y *around* W *suitably again, there exists a line bundle* $\mathcal{L}_Z$ *on* Z *such that* $\mathcal{L}\simeq\varphi_F^*\mathcal{L}_Z$ *holds.*

(5) *Every* $(K_X+\Delta)$*-negative extremal ray* R *with*

$$R\cap\overline{\mathrm{NE}}(X/Y;W)_{\mathrm{Nlc}(X,\Delta)}=\{0\}$$

is spanned by a (possibly singular) rational curve C *with*

$$0<-(K_X+\Delta)\cdot C\leq 2\dim X.$$

From now on, we further assume that (X, Δ) is log canonical, equivalently, $\mathrm{Nlc}(X, \Delta) = \emptyset$. *Then we have the following properties.*

(6) *Let H be an effective $\mathbb{R}$-Cartier $\mathbb{R}$-divisor on X such that $K_X+\Delta+H$ is π-nef over W and $(X, \Delta + H)$ is log canonical. Then, either $K_X + \Delta$ is also π-nef over W or there exists a $(K_X + \Delta)$-negative extremal ray R of $\overline{\mathrm{NE}}(X/Y; W)$ such that*

$$(K_X + \Delta + \lambda H) \cdot R = 0,$$

where

$$\lambda := \inf \{t \geq 0 \mid K_X + \Delta + tH \text{ is } \pi\text{-nef over } W\}.$$

Of course, $K_X + \Delta + \lambda H$ is π-nef over W.

Similarly, we have:

(7) *Let $\mathcal{H}$ be an $\mathbb{R}$-line bundle on X which is π-ample over W such that $K_X+\Delta+\mathcal{H}$ is π-nef over W. Then, either $K_X + \Delta$ is also π-nef over W or there exists a $(K_X + \Delta)$-negative extremal ray R of $\overline{\mathrm{NE}}(X/Y; W)$ such that*

$$(K_X + \Delta + \lambda \mathcal{H}) \cdot R = 0,$$

where

$$\lambda := \inf \{t \geq 0 \mid K_X + \Delta + t\mathcal{H} \text{ is } \pi\text{-nef over } W\}.$$

Note that $K_X + \Delta + \lambda \mathcal{H}$ is π-nef over W.

Remark 1.1.7. In Theorem 1.1.6, the proof of (5) needs Mori's bend and break method and (6) is an application of (5). On the other hand, (7) is an easy consequence of (3). Note that π-very ample line bundles do not always have global sections. Hence (7) is not a special case of (6). We need (7) in order to discuss the minimal model program of log canonical pairs with ample scaling for projective morphisms between complex analytic spaces.

For the minimal model program, the following theorem, which is a supplement to Theorem 1.1.6, may be useful (see [Fu13]).

Theorem 1.1.8. *Let (X, Δ) be a log canonical pair. Let $\pi\colon X \to Y$ be a projective morphism of complex analytic spaces and let W be a compact subset of Y such that the dimension of $N^1(X/Y; W)$ is finite. Suppose that $\pi\colon X \to Y$ is decomposed as*

$$\pi\colon X \xrightarrow{f} Y^\flat \xrightarrow{g} Y$$

such that $Y^\flat$ is projective over Y. Let $\mathcal{A}_{Y^\flat}$ be a g-ample line bundle on $Y^\flat$. Let R be a $(K_X + \Delta + (\dim X + 1)f^*\mathcal{A}_{Y^\flat})$-negative extremal ray of $\overline{\mathrm{NE}}(X/Y; W)$. Then R is a $(K_X+\Delta)$-negative extremal ray of $\overline{\mathrm{NE}}\left(X/Y^\flat; g^{-1}(W)\right)$, that is, $R \cdot f^*\mathcal{A}_{Y^\flat} = 0$.

We prove Theorem 1.1.8 as an application of the vanishing theorem for projective quasi-log schemes. We do not need Theorem 1.1.6 (5) for the proof of Theorem 1.1.8. We have the following result as an easy consequence of Theorem 1.1.8.

Corollary 1.1.9 (see [Fu10, Corollary 1.2]). *Let (X, Δ) be a log canonical pair. Let $\pi\colon X \to Y$ be a projective morphism of complex analytic spaces and let $\mathcal{A}$ be any π-ample line bundle on X. Then $K_X + \Delta + (\dim X + 1)\mathcal{A}$ is always nef over Y.*

We make an important remark on Theorem 1.1.6. By Remark 1.1.10, we see that the cone and contraction theorem of normal pairs holds for projective morphisms between *compact* analytic spaces.

Remark 1.1.10. Let $\pi\colon X \to Y$ be a projective morphism of complex analytic spaces and let W be a compact subset of Y. Then the dimension of $N^1(X/Y; W)$ is not always finite (see Example 4.5.1). In [Na2, Chapter II. 5.19. Lemma] (see Theorem 4.5.2), Noboru Nakayama proved that if

- W is a compact subset of Y such that $W \cap V$ has only finitely many connected components for any analytic subset V which is defined over an open neighborhood of W,

then the dimension of $N^1(X/Y; W)$ is finite. We note that the above assumption is satisfied in the following cases:

(i) W is a point of Y.

(ii) W is a compact semianalytic subset of Y.

(iii) $W = Y$ when Y is compact.

Case (i) is obvious. In Case (ii), $W \cap V$ is a compact semianalytic subset of Y. Thus we see that $W \cap V$ has only finitely many connected components (see, for example, [BieM1, Corollary 2.7 (2)]). In Case (iii), $W \cap V = V$ is a compact analytic subset of Y. Hence it has only finitely many connected components.

By Remark 1.1.10, we see that there are many compact subsets W of Y such that the dimension of $N^1(X/Y; W)$ is finite.

We note that we can formulate and prove the basepoint-free theorem for projective morphisms of complex analytic spaces as follows. In Theorem 1.1.11, L is only assumed to be π-nef over W, that is, $L|_{\pi^{-1}(w)}$ is nef in the usual sense for every $w \in W$. Equivalently, $L \cdot C \geq 0$ for every projective integral curve C on X such that $\pi(C)$ is a point of W. However, Theorem 1.1.11 claims that it is π-semiample over some open neighborhood of W.

Theorem 1.1.11 (Basepoint-free theorem: Theorem 4.2.1). *Let $\pi\colon X \to Y$ be a projective morphism of complex analytic spaces such that X is a normal complex variety and let W be a compact subset of Y. Let Δ be an effective $\mathbb{R}$-divisor on X such that $K_X + \Delta$ is $\mathbb{R}$-Cartier. Let L be a Cartier divisor on X which is π-nef over W. We assume that*

(i) *$aL - (K_X + \Delta)$ is π-ample over W for some positive real number a, and*

(ii) *there exists some open neighborhood of W over which the invertible sheaf $\mathcal{O}_{\mathrm{Nlc}(X,\Delta)}(mL)$ is $\pi|_{\mathrm{Nlc}(X,\Delta)}$-generated for every $m \gg 0$.*

Then there exists a relatively compact open neighborhood U of W such that $\mathcal{O}_X(mL)$ is π-generated over U for every $m \gg 0$.

In Theorem 1.1.11, W is only assumed to be a compact subset of Y. We do not need the assumption that $\dim_{\mathbb{R}} N^1(X/Y;W) < \infty$ holds. When (X,Δ) is log canonical, we will also prove the basepoint-free theorem for $\mathbb{R}$-Cartier divisors.

Theorem 1.1.12 (Basepoint-free theorem for $\mathbb{R}$-divisors, see Theorem 5.3.1)**.** *Let $\pi\colon X \to Y$ be a projective morphism of complex analytic spaces such that X is a normal complex variety and let W be a compact subset of Y such that the dimension of $N^1(X/Y;W)$ is finite. Let Δ be an effective $\mathbb{R}$-divisor on X such that (X,Δ) is log canonical. Let D be an $\mathbb{R}$-Cartier $\mathbb{R}$-divisor defined on some open neighborhood of $\pi^{-1}(W)$ such that D is π-nef over W. Assume that $aD-(K_X+\Delta)$ is π-ample over W for some positive real number a. Then there exists an open neighborhood U of W such that D is semiample over U.*

In Theorem 1.1.12, we have to assume that the dimension of $N^1(X/Y;W)$ is finite since we need the cone theorem for the proof of Theorem 5.3.1. In the proof of Theorems 1.1.6, 1.1.11, and so on, the following basic properties of log canonical centers play an important role.

Theorem 1.1.13 (Basic properties of log canonical centers: Theorem 3.4.1)**.** *Let (X,Δ) be a log canonical pair. Then the following properties hold.*

(1) *The number of log canonical centers of (X,Δ) is locally finite.*
(2) *The intersection of two log canonical centers is a union of some log canonical centers.*
(3) *Let $x \in X$ be any point such that (X,Δ) is log canonical but is not kawamata log terminal at x. Then there exists a unique minimal* (*with respect to the inclusion*) *log canonical center C_x passing through x. Moreover, C_x is normal at x.*

Theorem 1.1.13 is new for complex analytic log canonical pairs although it is well known when (X,Δ) is algebraic. It will be useful for the study of complex analytic log canonical singularities (see also [Fu14]).

Theorem 1.1.6 is a starting point of the minimal model program of log canonical pairs for projective morphisms between complex analytic spaces. We can formulate the minimal model theory of log canonical pairs for projective morphisms between complex analytic spaces by using Theorem 1.1.6 as in the algebraic case. For the recent developments of the minimal model program for complex analytic log canonical pairs, we strongly recommend the interested reader to see the following preprints: [EH1], [EH2], [Fu19], and [EH3]. In this book, we do not discuss the minimal model program for log canonical pairs. Here, we only explain one of the main goals of the minimal model theory for projective morphisms between complex analytic spaces.

Conjecture 1.1.14 (Finite generation)**.** *Let $\pi\colon X \to Y$ be a projective morphism of complex analytic spaces and let Δ be an effective $\mathbb{Q}$-divisor on X such that (X,Δ)*

is log canonical. Then

$$R(X/Y, K_X+\Delta) := \bigoplus_{m\in\mathbb{N}} \pi_*\mathcal{O}_X(\lfloor m(K_X+\Delta)\rfloor)$$

is a locally finitely generated graded $\mathcal{O}_Y$-algebra.

We note that in [Fu13] Conjecture 1.1.14 was already solved completely when (X, Δ) is kawamata log terminal. We also note that Conjecture 1.1.14 is still widely open even when $\pi\colon X\to Y$ is algebraic (see [FG2]). For some other conjectures, see Section 1.2.

The author first prepared a short manuscript which only explains how to modify arguments in [Fu4]. Unfortunately, however, it seemed to be hard to read. Hence he made great efforts to make this book as self-contained as possible except for the results established in [Fu15] (see also [Fu18] and [FF]). He sometimes repeats arguments in [Fu4] and [Fu6]. Thus, some parts of this book are very similar to those of [Fu4] and [Fu6].

Remark 1.1.15 (Quasi-log structures, see Section 6.1). By [Fu15, Theorems 1.1 and 1.2] (see Theorems 3.1.5 and 3.2.3), we can formulate and discuss quasi-log structures on complex analytic spaces (see [Fu6, Chapter 6]). Hence we can establish the cone and contraction theorem for highly singular complex analytic spaces. However, in this book, we will only discuss the cone and contraction theorem of normal pairs (see Theorem 1.1.6). This is because Theorem 1.1.6 is sufficient for many geometric applications and it is not so easy psychologically to treat reducible complex analytic spaces. We will describe the theory of quasi-log complex analytic spaces in [Fu16] (see also Sections 6.1, 6.2, and 6.3). By using the theory of quasi-log complex analytic spaces, we will establish the basepoint-free theorem of Reid–Fukuda type (see Theorem 6.2.1), the effective freeness (see Theorems 6.3.1 and 6.3.2), and the effective very ampleness (see Theorem 6.3.3) in [Fu16].

In this book, every complex analytic space is assumed to be *Hausdorff* and *second-countable*. An irreducible and reduced complex analytic space is called a *complex variety*. We will freely use the standard notation in [Fu4], [Fu6], [Fu13], and so on. We will also freely use the basic results on complex analytic geometry in [BăS] and [Fi]. For the minimal model program for projective morphisms between complex analytic spaces, see [Na1], [Na2], and [Fu13]. For the traditional framework of the minimal model program, see [KMM], [KM], and [Ma]. We note that $\mathbb{Z}$, $\mathbb{Q}$, and $\mathbb{R}$ denote the set of *integers*, *rational numbers*, and *real numbers*, respectively. We also note that $\mathbb{N}$ (resp. $\mathbb{Z}_{>0}$) is the set of *non-negative integers* (resp. *positive integers*).

1.2 Minimal model conjecture

In this section, we will discuss the minimal model conjecture for projective morphisms between complex analytic spaces. Here, we will not treat the minimal model program of complex analytic spaces systematically (see [Fu13]). We will only explain some conjectures related to the theory of minimal models in the complex analytic setting. For some related results, see Section 6.4. Let us start with the following important and difficult conjecture.

Conjecture 1.2.1. *Let $\pi\colon X \to Y$ be a projective morphism of complex analytic spaces, let (X, Δ) be a log canonical pair, and let W be a compact subset of Y. Assume that $K_X + \Delta$ is π-nef over W, that is, $(K_X + \Delta) \cdot C \geq 0$ holds for every projective integral curve C on X such that $\pi(C)$ is a point of W. Then $K_X + \Delta$ is nef over some open neighborhood U of W.*

Let us recall the definition of *small projective morphisms* in order to explain flips.

Definition 1.2.2 (Small projective morphisms)**.** Let $\varphi\colon X \to Z$ be a projective morphism of normal complex varieties. If the exceptional locus of φ has codimension ≥ 2, then φ is called *small.*

The existence of flips (Flip conjecture I) and the termination of flips (Flip conjecture II) are famous conjectures in the theory of minimal models. Fortunately, even in the complex analytic setting, the existence of flips is completely settled in a recent preprint [Fu19] with the aid of [EH1] and [EH2].

Theorem 1.2.3 (Flip conjecture I, [Fu19, Theorem 1.7])**.** *Let $\varphi\colon X \to Z$ be a small projective morphism of complex analytic spaces and let (X, Δ) be a log canonical pair. Assume that $-(K_X + \Delta)$ is φ-ample. Then we can construct the following commutative diagram:*

$$\begin{array}{ccc} (X, \Delta) & \overset{\phi}{\dashrightarrow} & (X^+, \Delta^+) \\ & {\scriptstyle\varphi}\searrow \quad \swarrow{\scriptstyle\varphi^+} & \\ & Z & \end{array}$$

where

(i) (X^+, Δ^+) *is log canonical,*

(ii) $\varphi^+\colon X^+ \to Z$ *is small projective, and*

(iii) $K_{X^+} + \Delta^+$ *is φ^+-ample, where Δ^+ is the strict transform of Δ on X^+.*

We call $\varphi^+\colon (X^+, \Delta^+) \to Z$ a $(K_X + \Delta)$-flip of φ. The exceptional locus of φ (resp. φ^+) is called the flipping *(resp.* flipped*) locus. We note that we frequently simply say that $\phi\colon X \dashrightarrow X^+$ is a flip if there is no danger of confusion. Furthermore, if Z is projective over a complex analytic space Y, then we say that $\phi\colon X \dashrightarrow X^+$ is a flip over Y.*

We note that the proof of Theorem 1.2.3 is not so easy even in the original algebraic setting (see [Has1]). We also note that Theorem 1.2.3 was established in [Fu13] when (X, Δ) is kawamata log terminal (see also [DHP] and [LM]). The termination of flips is one of the most difficult open problems in the theory of minimal models.

Conjecture 1.2.4 (Flip conjecture II)**.** *Let (X, Δ) be a log canonical pair and let*

$$(X, \Delta) =: (X_0, \Delta_0) \overset{\phi_0}{\dashrightarrow} (X_1, \Delta_1) \overset{\phi_1}{\dashrightarrow} \cdots \overset{\phi_{i-1}}{\dashrightarrow} (X_i, \Delta_i) \overset{\phi_i}{\dashrightarrow} \cdots$$

be a sequence of flips over a complex analytic space Y starting from (X, Δ). Let W be a compact subset of Y. Then, there exists some i_0 such that ϕ_i is an isomorphism over W for every $i \geq i_0$, that is, ϕ_i is an isomorphism over some open neighborhood U_i of W for every $i \geq i_0$.

Conjecture 1.2.4 is still widely open even when everything is algebraic. For some recent developments, see [EH3]. The abundance conjecture is also a very difficult open problem. It is also widely open even for projective varieties.

Conjecture 1.2.5 (Abundance conjecture)**.** *Let $\pi \colon X \to Y$ be a projective morphism of complex analytic spaces, let (X, Δ) be a log canonical pair, and let W be a compact subset of Y. Assume that $K_X + \Delta$ is π-nef over W. Then $K_X + \Delta$ is π-semiample over some open neighborhood U of W.*

Roughly speaking, in [Fu19], we essentially reduce Conjecture 1.2.5 to Conjecture 1.2.6, which is an important very special case of Conjecture 1.2.5. For the details, see [Fu19, Theorem 1.9 and Corollary 1.10].

Conjecture 1.2.6 (Abundance conjecture for projective log canonical pairs)**.** *Let (X, Δ) be a projective log canonical pair. Assume that $K_X+\Delta$ is nef. Then $K_X+\Delta$ is semiample.*

As is well known, Conjecture 1.2.6 seems to be very difficult and is still widely open. For the details of the abundance conjecture in the complex analytic setting, we recommend the reader to see [Fu19].

The finite generation of log canonical rings for log canonical pairs is an important open problem. For the details of Conjecture 1.2.7 for the case where $\pi \colon (X, \Delta) \to Y$ is algebraic, see [FG2].

Conjecture 1.2.7 (Finite generation, see Conjecture 1.1.14)**.** *Let $\pi \colon X \to Y$ be a projective morphism of complex analytic spaces and let Δ be an effective $\mathbb{Q}$-divisor on X such that (X, Δ) is log canonical. Then*

$$R(X/Y, K_X + \Delta) := \bigoplus_{m \in \mathbb{N}} \pi_* \mathcal{O}_X(\lfloor m(K_X + \Delta) \rfloor)$$

is a locally finitely generated graded $\mathcal{O}_Y$-algebra.

If (X, Δ) is a kawamata log terminal pair, then we know that Conjecture 1.2.7 holds true (see [Fu13] and Theorem 6.4.1). If (X, Δ) is log canonical but is not kawamata log terminal, then Conjecture 1.2.7 is open even when $\pi \colon (X, \Delta) \to Y$ is algebraic. For the details of the theory of minimal models of kawamata log terminal pairs for projective morphisms between complex analytic spaces, see [Fu13].

The final conjecture in this section is as follows.

Conjecture 1.2.8 (Finite generation for projective log canonical pairs)**.** *Let (X, Δ) be a projective log canonical pair such that Δ is a $\mathbb{Q}$-divisor. Then*

$$R(X, K_X + \Delta) := \bigoplus_{m \in \mathbb{N}} H^0(X, \mathcal{O}_X(\lfloor m(K_X + \Delta) \rfloor))$$

is a finitely generated $\mathbb{C}$-algebra.

By combining [FG2] with [EH3], we see that if Conjecture 1.2.8 holds for all projective log canonical pairs then Conjecture 1.2.7 also holds true. Anyway, almost all conjectures of the minimal model theory for projective morphisms between complex analytic spaces follow from the original conjectures for projective varieties.

We close this section with a historical comments.

Remark 1.2.9 (Historical comments). In [Fu15], we established the necessary vanishing theorems for the minimal model theory of log canonical pairs in the complex analytic setting (see Chapter 3). Then, in this book, we proved the cone and contraction theorem of log canonical pairs for projective morphisms between complex analytic spaces (see Chapter 4). After that, in [EH1] and [EH2], Enokizono and Hashizume discussed the minimal model program for log canonical pairs in the complex analytic setting. Then, in [Fu19], which is an analytic generalization of [Fu1] and [FG1], we studied the abundance conjecture for projective morphisms of complex analytic spaces. Finally, based on [Fu19], Enokizono and Hashizume strengthened some results of [EH2] in [EH3].

1.3 Algebraic vs Analytic

In this section, we will see various differences between algebraic varieties and complex analytic spaces for the reader's convenience. The reader can find some interesting examples in [Sh, Chapter 8, Sections 1.4, 3.2, and 3.3].

Let X be an algebraic variety. Then we can always take a completion $X \hookrightarrow \overline{X}$, that is, $\overline{X}$ is a complete algebraic variety which contains X as a Zariski open dense subset of $\overline{X}$. If $\overline{X}_1$ and $\overline{X}_2$ are two different completions of X, then it is obvious that $\overline{X}_1$ is birationally equivalent to $\overline{X}_2$ by definition. Unfortunately, a complex analytic space is not always compactifiable. Moreover, two different compactifications of a given complex manifold are not necessarily bimeromorphically equivalent. Let us recall the following famous example due to Serre.

Example 1.3.1 (Serre, see [Har, Chapter VII, Example 3.2])**.** Let C be an elliptic curve and let $\mathcal{E}$ be the rank two vector bundle on C which is defined by the unique non-splitting extension

$$0 \to \mathcal{O}_C \to \mathcal{E} \to \mathcal{O}_C \to 0$$

We consider the ruled surface $X := \mathbb{P}_C(\mathcal{E})$. On that surface there exists a unique section

$$D := \mathbb{P}_C(\mathcal{O}_C) \subset X$$

of $X \to C$ such that $\mathcal{O}_D(D) \simeq \mathcal{O}_D$ holds and that $\mathcal{O}_X(D) \simeq \mathcal{O}_{\mathbb{P}_C(\mathcal{E})}(1)$ is a nef line bundle on X. We put $U := X \setminus D$. Then U is obviously a quasi-projective algebraic surface. We can easily check that U is not affine. Let U^{an} be the complex analytic space naturally associated to U. Then we can check that U^{an} is analytically isomorphic to $\mathbb{C}^\times \times \mathbb{C}^\times$, where $\mathbb{C}^\times = \mathbb{C} \setminus \{0\}$. In particular, U^{an} is a Stein manifold. This example shows that the complex analytic space $\mathbb{C}^\times \times \mathbb{C}^\times$ has two different compactifications as follows:

$$X = \mathbb{P}_C(\mathcal{E}) \hookleftarrow \mathbb{C}^\times \times \mathbb{C}^\times \hookrightarrow \mathbb{P}^1 \times \mathbb{P}^1.$$

Note that X is not bimeromorphically equivalent to $\mathbb{P}^1 \times \mathbb{P}^1$.

Here we recall the precise definition of *bimeromorphic maps* for the sake of completeness.

Definition 1.3.2 (Bimeromorphic maps)**.** A *meromorphic map* $f\colon X \dashrightarrow Y$ of complex analytic varieties is defined by the graph $\Gamma_f \subset X \times Y$ such that Γ_f is a subvariety of $X \times Y$ and that the first projection is proper and is an isomorphism over a Zariski open dense subset of X. Note that a *Zariski open subset* is the complement of an analytic subset. If further the second projection $\Gamma_f \to Y$ is proper and is an isomorphism over a Zariski open dense subset of Y, then $f\colon X \dashrightarrow Y$ is called a *bimeromorphic map*. If there exists a bimeromprhic map between X and Y, then we say that X is *bimeromorphically equivalent* to Y.

In Example 1.3.1, the graph of the analytic isomorphism $U \xrightarrow{\sim} \mathbb{C}^\times \times \mathbb{C}^\times$ can not be extended to a closed subvariety of $X \times (\mathbb{P}^1 \times \mathbb{P}^1)$. This is the reason why X is not bimeromorphically equivalent to $\mathbb{P}^1 \times \mathbb{P}^1$.

Let $\mathcal{L}$ be a line bundle on an algebraic variety X. Then we can always find a Cartier divisor D on X such that $\mathcal{L} \simeq \mathcal{O}_X(D)$. This property does not always hold for complex analytic varieties. A simplest example, which is due to Siegel, is as follows.

Example 1.3.3 (Complex tori)**.** Let T be a two-dimensional complex torus with a period matrix

$$\Omega = \begin{pmatrix} 1 & 0 & \sqrt{-2} & \sqrt{-5} \\ 0 & 1 & \sqrt{-3} & \sqrt{-7} \end{pmatrix}.$$

Then it is well known that it has no non-constant meromorphic functions. Of course, the algebraic dimension of T is zero and there are no one-dimensional closed subvarieties on T. For the details, see, for example, [Sh, Example 8.4], where some topological arguments are used. We note that T is Kähler.

Note that there are many compact complex manifolds which have no non-constant meromorphic functions.

Example 1.3.4 (Hopf surfaces)**.** We put $W := \mathbb{C}^2 \setminus \{(0,0)\}$. Let G be the group of transformations

$$(z_1, z_2) \mapsto (2^k z_1, 3^k z_2)$$

of W with $k \in \mathbb{Z}$. Then we can easily check that $X := W/G$ is a compact complex surface which has no non-constant meromorphic functions. It is well known that X is diffeomorphic to $S^1 \times S^3$ and is not Kähler. More precisely, we have $b_1(X) = 1$ and $b_2(X) = 0$, where $b_i(X)$ denotes the ith Betti number of X. For the details, see, for example, [Kod, Example 2.9].

In this book, we are mainly interested in non-compact complex analytic spaces. We have to take care of the following simple example. For a related topic, see [Fu14].

Example 1.3.5 ($\mathbb{Q}$-Cartier $\mathbb{Q}$-divisors). Let $\{P_n\}_{n\in\mathbb{Z}_{>0}}$ be a set of mutually distinct discrete points of $X := \mathbb{C}$. Then

$$\Delta := \sum_{n\in\mathbb{Z}_{>0}} \frac{n-1}{n} P_n$$

is a $\mathbb{Q}$-Cartier $\mathbb{Q}$-divisor on a complex manifold X. In this case, there is no positive integer m such that $m\Delta$ is Cartier. We note that (X, Δ) is kawamata log terminal. However, $(X, t\Delta)$ is not log canonical for every $t > 1$.

If we apply Theorem 1.1.11, which is the basepoint-free theorem, to the usual algebraic setting, we have the following basepoint-free theorem, which is related to Conjecture 1.2.1.

Theorem 1.3.6. *Let $\pi\colon X \to Y$ be a projective morphism of quasi-projective algebraic varieties such that (X, Δ) is log canonical. Let $P \in Y$ be a closed point. Let L be a Cartier divisor on X such that $L \cdot C \geq 0$ holds for every projective integral curve C on X with $\pi(C) = P$. Assume that $aL - (K_X + \Delta)$ is π-ample for some positive real number a. Then there exists a Zariski open neighborhood U of P such that $\mathcal{O}_X(mL)$ is π-generated over U for every $m \gg 0$. In particular, L is π-nef over U.*

We prove Theorem 1.3.6 for the sake of completeness.

Proof of Theorem 1.3.6. By Theorem 1.1.11, we can take positive integers m_1 and m_2 such that $(m_1, m_2) = 1$ and that $\mathcal{O}_X(m_1L)$ and $\mathcal{O}_X(m_2L)$ are π-generated over an open neighborhood U' of P. We note that

$$\mathcal{S}_i := \pi\left(\mathrm{Supp}(\pi^*\pi_*\mathcal{O}_X(m_iL) \to \mathcal{O}_X(m_iL))\right)$$

is a closed analytic subset of Y which is disjoint from U' for $i = 1, 2$. We put

$$U := Y \setminus \bigcup_{i=1,2} \mathcal{S}_i.$$

Then we can easily see that this U is a desired Zariski open neighborhood of P. □

Note that L is not assumed to be π-nef in Theorem 1.3.6. We only assumed that L is π-nef over P. Therefore, Theorem 1.3.6 is new even when $\pi\colon (X, \Delta) \to Y$ is algebraic. The final example in this section is more or less well known.

Example 1.3.7 (see, for example, [Fu2, Example 3.2.4]). We consider

$$X := \{(x, y, z, w) \in \mathbb{C}^4 \mid xy + zw + z^3 + w^3 = 0\}.$$

Then X is a $\mathbb{Q}$-factorial affine algebraic variety, that is, every prime divisor on X is $\mathbb{Q}$-Cartier. More precisely, the coordinate ring

$$\mathbb{C}[x, y, z, w]/(xy + zw + z^3 + w^3)$$

of X is a UFD. Let us consider a germ of X^{an} around the origin, where X^{an} is the complex analytic space associated to X. It is easy to see that X^{an} is local analytically isomorphic to $(xy - uv = 0) \subset \mathbb{C}^4$ around the origin. The two divisors $(x = u = 0)$ and $(y = v = 0)$ intersect at a single point. Hence, on a germ of X^{an}, there exists a prime divisor which is not $\mathbb{Q}$-Cartier at the origin.

By Example 1.3.7, we see that the notion of $\mathbb{Q}$-factoriality is subtle for normal complex varieties. For the details of the strong $\mathbb{Q}$-factoriality necessary for the minimal model program of complex analytic spaces, see Section 5.4.

1.4 Precise contents

In this section, we will describe the contents of each chapter more precisely for the reader's convenience.

Chapter 2: Chapter 2 is a preliminary chapter. In Section 2.1, we collect some necessary definitions and results for the reader's convenience. Since we have to work in the complex analytic setting, some of them become much more subtle than the usual ones in the algebraic setting. In Section 2.2, we collect some basic properties of relatively nef and relatively ample $\mathbb{R}$-line bundles for the sake of completeness. They are indispensable in subsequent sections. We note that we have to treat $\mathbb{R}$-line bundles in the theory of minimal models. In Section 2.3, we define non-lc ideal sheaves in the complex analytic setting and prove some elementary lemmas.

Chapter 3: Chapter 3 is devoted to our new vanishing theorems in the complex analytic setting established in [Fu15] and their direct consequences. In Section 3.1, we quickly recall the main result of [Fu15]. Note that the proof of the main result in [Fu15] depends on Saito's theory of mixed Hodge modules and Takegoshi's generalization of Kollár's torsion-free and vanishing theorem. However, we can make the main result of [Fu15] free from the theory of mixed Hodge modules by [FF]. Section 3.2 is a short section, where we treat vanishing theorems of Reid–Fukuda type. We note that we do not need them in this book. In Section 3.3, we prove some vanishing theorems for normal pairs as applications of the vanishing result explained in Section 3.1. In Section 3.4, we establish the basic properties of log canonical centers. They are new and very important in the theory of minimal models in the complex analytic setting.

Chapter 4: Chapter 4 is the main part of this book. Here we establish the fundamental theorems of the theory of minimal models for projective morphisms between complex analytic spaces. In Section 4.1, we prove the non-vanishing theorem in the complex analytic setting with the aid of the theory of quasi-log schemes. Note that Lemma 4.1.3 is new and will be useful for the study of quasi-log structures. In Section 4.2, we establish the basepoint-free theorem for normal pairs in the complex analytic setting by using the non-vanishing theorem proved in Section 4.1. It is well known and is not difficult to prove for kawamata log terminal pairs. In Sections 4.3, we prove the rationality theorem for normal pairs in the complex analytic setting. The proof is essentially the same as the one for algebraic varieties explained in [Fu4]. In Section 4.4, we define Kleiman–Mori cones for projective morphisms of complex analytic spaces. Then we prove Kleiman's ampleness criterion in the complex analytic setting. In Section 4.5, we briefly explain Nakayama's finiteness without proof for the reader's convenience. Note that in this book we do not need it except in the proof of Corollary 1.1.9. Then, in Section 4.6, we prove the cone theorem for normal pairs in the complex analytic setting. The results in Section 4.6 are easy consequences of the basepoint-free theorem in Section 4.2 and the rationality theorem in Section 4.3. In the final section: Section 4.7, we prove Theorem 1.1.8 as an easy application of the vanishing theorem for projective quasi-log schemes.

Chapter 5: In Chapter 5, we treat some supplementary results. In Section 5.1, we discuss lengths of extremal rational curves. The result in Section 5.1 seems to be indispensable for the minimal model program with scaling. Here, we use the framework of quasi-log schemes. In Section 5.2, we discuss Shokurov's polytopes and some applications. The results in this section are well known and have already played an important role in the usual algebraic setting. In Section 5.3, we prove the basepoint-free theorem of log canonical pairs for $\mathbb{R}$-Cartier divisors. It can be seen as an application of the cone theorem. In Section 5.4, we discuss strong $\mathbb{Q}$-factoriality. It is indispensable for the minimal model program of complex analytic spaces. This is because a divisorial sheaf on a normal complex variety does not necessarily come from a Weil divisor. In Section 5.5, which is the final section of this chapter, we prove the main result of this book, that is, the cone and contraction theorem of normal pairs for projective morphisms between complex analytic spaces: Theorem 1.1.6.

Chapter 6: Chapter 6 is devoted to some advanced topics. In Section 6.1, we quickly review the theory of quasi-log complex analytic spaces without proof. The reader can find all the details in [Fu16]. In Section 6.2, we explain the basepoint-free theorem of Reid–Fukuda type for log canonical pairs in the complex analytic setting (see Theorem 6.2.1). The proof of Theorem 6.2.1 needs the framework of quasi-log complex analytic spaces and is much harder than the one for the usual basepoint-free theorem (see Theorem 4.2.1). In Section 6.3, we review the effective freeness and very ampleness for log canonical pairs in the complex analytic setting. In Section 6.4, we explain some results on the minimal model program for projective morphisms between complex analytic spaces established in [Fu13].

Chapter 7: This chapter is an appendix, where we correct some minor mistakes in the author's book:

> [Fu6] O. Fujino, *Foundations of the minimal model program*, MSJ Memoirs, **35**. Mathematical Society of Japan, Tokyo, 2017

for the reader's convenience.

Chapter 2
Preliminaries

In this chapter, we will collect basic definitions and results necessary for this book. In Section 2.1, we will discuss singularities of pairs for complex analytic spaces and explain various basic properties. In Section 2.2, we will treat relatively ample and relatively nef $\mathbb{R}$-line bundles in the complex analytic setting for the reader's convenience. In Section 2.3, we will introduce the notion of non-lc ideal sheaves for normal complex varieties and prove some basic properties.

2.1 Singularities of pairs

In this section, we will collect basic definitions and results necessary for this book. For the details, see [Fu4], [Fu6], [Fu13], and so on. Since we are working in the complex analytic setting, some of them become subtle.

Let us start with the definition of *singularities of pairs*, which is indispensable in the theory of minimal models.

2.1.1 (Singularities of pairs, log canonical centers, and non-lc loci). We consider a normal complex variety X. Let X_{sm} denote the smooth locus of X. Then the *dualizing sheaf* ω_X of X is the unique reflexive sheaf whose restriction to X_{sm} is isomorphic to the sheaf $\Omega^n_{X_{\mathrm{sm}}}$, where $n = \dim X$. Note that ω_X is sometimes called the *canonical sheaf* of X. Let Δ be an $\mathbb{R}$-divisor on X, that is, Δ is a locally finite $\mathbb{R}$-linear combination of prime divisors (see 2.1.9) on X. We say that $K_X + \Delta$ is $\mathbb{R}$-*Cartier* at $x \in X$ if there exist an open neighborhood U_x of x and a Weil divisor K_{U_x} on U_x with $\mathcal{O}_{U_x}(K_{U_x}) \simeq \omega_X|_{U_x}$ such that $K_{U_x} + \Delta|_{U_x}$ is $\mathbb{R}$-Cartier, that is, $K_{U_x} + \Delta|_{U_x}$ is a finite $\mathbb{R}$-linear combination of Cartier divisors on U_x. For any subset L of X, we say that $K_X + \Delta$ is $\mathbb{R}$-Cartier at L if it is $\mathbb{R}$-Cartier at any point x of L. We simply say that $K_X + \Delta$ is $\mathbb{R}$-Cartier when $K_X + \Delta$ is $\mathbb{R}$-Cartier at any point $x \in X$. Unfortunately, however, we can not always define K_X globally with $\mathcal{O}_X(K_X) \simeq \omega_X$. In general, it only exists locally on X. We usually use the symbol K_X as a formal divisor class with an isomorphism $\mathcal{O}_X(K_X) \simeq \omega_X$ and call it the *canonical divisor* of X if there is no danger of confusion.

Let $f\colon Y \to X$ be a proper bimeromorphic morphism from a normal complex variety Y. Suppose that $K_X + \Delta$ is $\mathbb{R}$-Cartier in the above sense. We take a small Stein open subset U of X where $K_U + \Delta|_U$ is a well-defined $\mathbb{R}$-Cartier $\mathbb{R}$-divisor on U. In this situation, we can define $K_{f^{-1}(U)}$ and K_U such that $f_*K_{f^{-1}(U)} = K_U$. Then we can write

$$K_{f^{-1}(U)} = f^*(K_U + \Delta|_U) + E_U$$

as usual. Note that E_U is a well-defined $\mathbb{R}$-divisor on $f^{-1}(U)$ such that $f_*E_U = \Delta|_U$. Then we have the following formula

$$K_Y = f^*(K_X + \Delta) + \sum_E a(E, X, \Delta)E$$

as in the algebraic case. We note that $\sum_E a(E, X, \Delta)E$ is a globally well-defined $\mathbb{R}$-divisor on Y such that

$$\left(\sum_E a(E, X, \Delta)E\right)\Bigg|_{f^{-1}(U)} = E_U$$

although K_X and K_Y are well defined only locally.

If Δ is a *boundary* $\mathbb{R}$*-divisor* (see 2.1.9), that is, all the coefficients of Δ are in $[0, 1] \cap \mathbb{R}$, and $a(E, X, \Delta) \geq -1$ holds for any $f : Y \to X$ and every f-exceptional divisor E, then (X, Δ) is called a *log canonical* pair. If (X, Δ) is log canonical and $a(E, X, \Delta) > -1$ for any $f : Y \to X$ and every f-exceptional divisor E, then (X, Δ) is called a *purely log terminal* pair. If (X, Δ) is purely log terminal and $\lfloor \Delta \rfloor = 0$, that is, the coefficients of Δ are in $[0, 1) \cap \mathbb{R}$, then (X, Δ) is called a *kawamata log terminal* pair. When $\Delta = 0$ and $a(E, X, 0) \geq 0$ (resp. > 0) for any $f : Y \to X$ and every f-exceptional divisor E, we simply say that X has only *canonical singularities* (resp. *terminal singularities*). In this book, we will only use log canonical pairs and kawamata log terminal pairs.

More generally, let X be a normal complex variety and let Δ be an effective $\mathbb{R}$-divisor on X. We say that (X, Δ) is *log canonical* (resp. *kawamata log terminal*) *at* $x \in X$ if there exists an open neighborhood U_x of x such that $(U_x, \Delta|_{U_x})$ is a log canonical pair (resp. kawamata log terminal pair). Let L be any subset of X. We say that (X, Δ) is *log canonical* (resp. *kawamata log terminal*) *at* L if (X, Δ) is log canonical (resp. kawamata log terminal) at any point x of L. We note that (X, Δ) is log canonical (resp. kawamata log terminal) in the above sense if and only if (X, Δ) is log canonical (resp. kawamata log terminal) at any point x of X.

Let X be a normal complex variety and let Δ be an effective $\mathbb{R}$-divisor on X such that $K_X + \Delta$ is $\mathbb{R}$-Cartier. The image of E with $a(E, X, \Delta) = -1$ for some $f : Y \to X$ such that (X, Δ) is log canonical around general points of $f(E)$ is called a *log canonical center* of (X, Δ). The *non-lc locus* of (X, Δ), denoted by $\mathrm{Nlc}(X, \Delta)$, is the smallest closed subset Z of X such that the complement $(X \setminus Z, \Delta|_{X\setminus Z})$ is log canonical. We can define a natural complex analytic space structure on $\mathrm{Nlc}(X, \Delta)$ by the non-lc ideal sheaf $\mathcal{J}_{\mathrm{NLC}}(X, \Delta)$ of (X, Δ). For the definition of $\mathcal{J}_{\mathrm{NLC}}(X, \Delta)$, see Section 2.3 below.

The above definition is compatible with the usual definition for algebraic varieties.

Remark 2.1.2 (cf. [Ma, Proposition 4-4-4]). Let (X, Δ) be a pair consisting of a normal algebraic variety X and an effective $\mathbb{R}$-divisor on X such that $K_X + \Delta$ is $\mathbb{R}$-Cartier. Then (X, Δ) is kawamata log terminal (resp. log canonical) in the usual sense (see [Fu4], [Fu6], and so on) if and only if $(X^{\mathrm{an}}, \Delta^{\mathrm{an}})$ is kawamata log terminal (resp. log canonical) in the above sense, where X^{an} is the complex analytic space naturally associated to X and let Δ^{an} be the $\mathbb{R}$-divisor on X^{an} associated to Δ.

The following lemma is well known for algebraic varieties.

Lemma 2.1.3. *Let X be a normal complex variety and let Δ be an effective $\mathbb{R}$-divisor on X such that $K_X+\Delta$ is $\mathbb{R}$-Cartier. Let P be a point of X and let D_i be an effective Cartier divisor on X with $P \in \operatorname{Supp} D_i$ for every i. If $\left(X, \Delta+\sum_{i=1}^{k} D_i\right)$ is log canonical at P, then $k \leq \dim X$ holds.*

We omit the proof of Lemma 2.1.3 here since the usual proof for algebraic varieties can work without any changes (see, for example, [Fu4, Lemma 13.2]). We will use Lemma 2.1.3 in order to create a new log canonical center.

In this book, we sometimes implicitly use Serre's GAGA.

2.1.4 (Serre's GAGA)**.** Let $\pi\colon X \to Y$ be a projective morphism of complex analytic spaces and let F be a fiber of $\pi\colon X \to Y$. Then F is projective. Hence we can apply various results of projective schemes to F with the aid of Serre's GAGA (see [Se]).

In the theory of minimal models, we need the notion of $\mathbb{R}$-*line bundles* and $\mathbb{Q}$-*line bundles*.

2.1.5 (Line bundles, $\mathbb{R}$-line bundles, and $\mathbb{Q}$-line bundles)**.** Let X be a complex analytic space and let $\operatorname{Pic}(X)$ denote the group of line bundles on X, that is, the *Picard group* of X. An element of $\operatorname{Pic}(X) \otimes_{\mathbb{Z}} \mathbb{R}$ (resp. $\operatorname{Pic}(X) \otimes_{\mathbb{Z}} \mathbb{Q}$) is called an $\mathbb{R}$-*line bundle* (resp. a $\mathbb{Q}$-*line bundle*) on X. In this book, we usually write the group law of $\operatorname{Pic}(X) \otimes_{\mathbb{Z}} \mathbb{R}$ additively for simplicity of notation. Hence we sometimes use $m\mathcal{L}$ to denote $\mathcal{L}^{\otimes m}$ for $\mathcal{L} \in \operatorname{Pic}(X)$ and $m \in \mathbb{Z}$.

We also need the notion of $\mathbb{R}$-*divisors* and $\mathbb{Q}$-*divisors*.

2.1.6 (Divisors, $\mathbb{R}$-divisors, and $\mathbb{Q}$-divisors)**.** Let X be a reduced equidimensional complex analytic space. A *prime divisor* on X is an irreducible and reduced closed analytic subspace of codimension one. An $\mathbb{R}$-*divisor* D on X is a formal sum

$$D = \sum_i a_i D_i,$$

where D_i is a prime divisor on X with $D_i \neq D_j$ for $i \neq j$, $a_i \in \mathbb{R}$ for every i, and the *support*

$$\operatorname{Supp} D := \bigcup_{a_i \neq 0} D_i$$

is a closed analytic subset of X. In other words, the formal sum $\sum_i a_i D_i$ is locally finite. If $a_i \in \mathbb{Z}$ (resp. $a_i \in \mathbb{Q}$) for every i, then D is called a *divisor* (resp. $\mathbb{Q}$-*divisor*) on X. Note that a divisor is sometimes called an *integral Weil divisor* in order to emphasize the condition that $a_i \in \mathbb{Z}$ for every i. If $0 \leq a_i \leq 1$ (resp. $a_i \leq 1$) holds for every i, then an $\mathbb{R}$-divisor D is called a *boundary* (resp. *subboundary*) $\mathbb{R}$-divisor.

Let $D=\sum_i a_i D_i$ be an $\mathbb{R}$-divisor on X such that D_i is a prime divisor for every i with $D_i \neq D_j$ for $i \neq j$. The *round-down* $\lfloor D \rfloor$ of D is defined to be the divisor

$$\lfloor D \rfloor = \sum_i \lfloor a_i \rfloor D_i,$$

where $\lfloor x \rfloor$ is the integer defined by $x-1<\lfloor x \rfloor \leq x$ for every real number x. The *round-up* and the *fractional part* of D are defined to be

$$\lceil D \rceil := -\lfloor -D \rfloor, \quad \text{and} \quad \{D\} := D - \lfloor D \rfloor,$$

respectively. We put

$$D^{=1} := \sum_{a_i=1} D_i, \quad D^{<1} := \sum_{a_i<1} a_i D_i, \quad \text{and} \quad D^{>1} := \sum_{a_i>1} a_i D_i.$$

Let D be an $\mathbb{R}$-divisor on X and let x be a point of X. If D is written as a finite $\mathbb{R}$-linear (resp. $\mathbb{Q}$-linear) combination of Cartier divisors on some open neighborhood of x, then D is said to be *$\mathbb{R}$-Cartier at x* (resp. *$\mathbb{Q}$-Cartier at x*). If D is $\mathbb{R}$-Cartier (resp. $\mathbb{Q}$-Cartier) at x for every $x \in X$, then D is said to be *$\mathbb{R}$-Cartier* (resp. *$\mathbb{Q}$-Cartier*). More generally, for any subset L of X, if D is $\mathbb{R}$-Cartier (resp. $\mathbb{Q}$-Cartier) at x for every $x \in L$, then D is said to be *$\mathbb{R}$-Cartier* (resp. *$\mathbb{Q}$-Cartier*) *at L*. Note that a $\mathbb{Q}$-Cartier $\mathbb{R}$-divisor D is automatically a $\mathbb{Q}$-Cartier $\mathbb{Q}$-divisor by definition. If D is a finite $\mathbb{R}$-linear (resp. $\mathbb{Q}$-linear) combination of Cartier divisors on X, then we sometimes say that D is a *globally $\mathbb{R}$-Cartier $\mathbb{R}$-divisor* (resp. *globally $\mathbb{Q}$-Cartier $\mathbb{Q}$-divisor*).

Two $\mathbb{R}$-divisors D_1 and D_2 are said to be *linearly equivalent* if $D_1 - D_2$ is a principal Cartier divisor. The linear equivalence is denoted by $D_1 \sim D_2$. Two $\mathbb{R}$-divisors D_1 and D_2 are said to be *$\mathbb{R}$-linearly equivalent* (resp. *$\mathbb{Q}$-linearly equivalent*) if $D_1 - D_2$ is a *finite* $\mathbb{R}$-linear (resp. $\mathbb{Q}$-linear) combination of principal Cartier divisors. When D_1 is $\mathbb{R}$-linearly (resp. $\mathbb{Q}$-linearly) equivalent to D_2, we write $D_1 \sim_{\mathbb{R}} D_2$ (resp. $D_1 \sim_{\mathbb{Q}} D_2$).

Example 2.1.7. Let X be a non-compact Riemann surface and let $\{P_k\}_{k=1}^{\infty}$ be a set of mutually distinct discrete points of X. We put $D := \sum_{k=1}^{\infty} \frac{1}{k} P_k$. Then D is obviously a $\mathbb{Q}$-Cartier $\mathbb{Q}$-divisor on X. However, D is not a finite $\mathbb{Q}$-linear combination of Cartier divisors on X.

We note that in this book we can almost always assume that $\operatorname{Supp} D$ has only finitely many irreducible components.

2.1.8 (Hybrids of $\mathbb{R}$-line bundles and $\mathbb{R}$-Cartier divisors). In this book, we usually treat hybrids of $\mathbb{R}$-line bundles and $\mathbb{R}$-Cartier divisors.

Let $\pi\colon X \to Y$ be a projective morphism between complex analytic spaces and let W be a compact subset of Y. Let A and B be $\mathbb{R}$-Cartier divisors on X and let $\mathcal{L}$ and $\mathcal{M}$ be $\mathbb{R}$-line bundles on X.

We sometimes say that

$$\mathcal{L} + A \sim_{\mathbb{R}} \mathcal{M} + B$$

holds over some open neighborhood U of W. This means:

(i) We implicitly assume that $A|_{\pi^{-1}(U)}$ and $B|_{\pi^{-1}(U)}$ are finite $\mathbb{R}$-linear combinations of Cartier divisors on $\pi^{-1}(U)$. Thus we can obtain $\mathbb{R}$-line bundles $\mathcal{A}$ and $\mathcal{B}$ naturally associated to $A|_{\pi^{-1}(U)}$ and $B|_{\pi^{-1}(U)}$, respectively.

(ii) In $\mathrm{Pic}(\pi^{-1}(U)) \otimes_{\mathbb{Z}} \mathbb{R}$, the following equality

$$\mathcal{L}|_{\pi^{-1}(U)} + \mathcal{A} = \mathcal{M}|_{\pi^{-1}(U)} + \mathcal{B}$$

holds.

If X is a normal complex variety and U is a relatively compact open subset of Y, then $A|_{\pi^{-1}(U)}$ and $B|_{\pi^{-1}(U)}$ are automatically finite $\mathbb{R}$-linear combinations of Cartier divisors on $\pi^{-1}(U)$. Therefore, (i) is harmless for applications.

Similarly, we say that $\mathcal{L} + A$ is π-ample over some open neighborhood U of W if $A|_{\pi^{-1}(U)}$ is a finite $\mathbb{R}$-linear combination of Cartier divisors on $\pi^{-1}(U)$, $\mathcal{A}$ is the $\mathbb{R}$-line bundle naturally associated to $A|_{\pi^{-1}(U)}$, and $\mathcal{L}|_{\pi^{-1}(U)} + \mathcal{A}$ is π-ample over U, that is, $\mathcal{L}|_{\pi^{-1}(U)} + \mathcal{A}$ is a finite positive $\mathbb{R}$-linear combination of π-ample line bundles on $\pi^{-1}(U)$.

2.1.9. Let $\pi\colon X \to Y$ be a projective morphism of complex analytic spaces such that X is a normal complex variety, and let Δ be an $\mathbb{R}$-divisor on X such that $K_X + \Delta$ is $\mathbb{R}$-Cartier. Let y be an arbitrary point of Y and let U_y be any relatively compact Stein open neighborhood of $y \in Y$. In this case, we can always find a Weil divisor $K_{\pi^{-1}(U_y)}$ on $\pi^{-1}(U_y)$ such that $\mathcal{O}_{\pi^{-1}(U_y)}(K_{\pi^{-1}(U_y)}) \simeq \omega_{\pi^{-1}(U_y)}$ holds since π is projective and U_y is Stein. Since U_y is relatively compact, $\operatorname{Supp} \Delta|_{\pi^{-1}(U_y)}$ has only finitely many irreducible components. If necessary, we replace U_y with a smaller relatively compact Stein open neighborhood of $y \in Y$. Then we can easily see that $K_{\pi^{-1}(U_y)} + \Delta|_{\pi^{-1}(U_y)}$ is a globally $\mathbb{R}$-Cartier $\mathbb{R}$-divisor on $\pi^{-1}(U_y)$. Moreover, for any $\mathbb{R}$-line bundle $\mathcal{L}$ on X, we can take a globally $\mathbb{R}$-Cartier $\mathbb{R}$-divisor L on $\pi^{-1}(U_y)$ such that $\mathcal{L}|_{\pi^{-1}(U_y)}$ is the $\mathbb{R}$-line bundle naturally associated to L.

In the theory of minimal models, we often use the following formulation. We will repeatedly use it in subsequent sections.

2.1.10. Let X be a normal complex variety. A real vector space spanned by the prime divisors on X is denoted by $\mathrm{WDiv}_{\mathbb{R}}(X)$, which has a canonical basis given by the prime divisors. Let D be an element of $\mathrm{WDiv}_{\mathbb{R}}(X)$. Then the sup norm of D with respect to this basis is denoted by $\|D\|$. Note that an $\mathbb{R}$-divisor D on X is an element of $\mathrm{WDiv}_{\mathbb{R}}(X)$ if and only if $\operatorname{Supp} D$ has only finitely many irreducible components.

Let V be a finite-dimensional affine subspace of $\mathrm{WDiv}_{\mathbb{R}}(X)$, which is defined over the rationals. We put

$$\mathcal{R}(V; x) := \{\Delta \in V \mid K_X + \Delta \text{ is } \mathbb{R}\text{-Cartier at } x\}.$$

It is obvious that $\mathcal{R}(V; x)$ is an affine subspace of V. We take an arbitrary element Δ of $\mathcal{R}(V; x)$. Then $K_X + \Delta$ is $\mathbb{R}$-Cartier at x by definition. Therefore, there exists a small open neighborhood U_x of x such that

$$K_{U_x} + \Delta|_{U_x} = \sum_{i=1}^{k} a_i D_i,$$

where D_i is a Cartier divisor on U_x and a_i is a real number for every i. By this description, we can easily see that there exists an affine subspace $\mathcal{T}$ of V defined over the rationals such that $\Delta \in \mathcal{T} \subset \mathcal{R}(V;x)$. Hence $\mathcal{R}(V;x)$ itself is an affine subspace of V defined over the rationals. Let L be a compact subset of X. We put

$$\mathcal{R}(V;L) := \{\Delta \in V \mid K_X + \Delta \text{ is } \mathbb{R}\text{-Cartier at } L\}.$$

Then the following equality

$$\mathcal{R}(V;L) = \bigcap_{x \in L} \mathcal{R}(V;x)$$

obviously holds. Therefore, $\mathcal{R}(V;L)$ is an affine subspace of V defined over the rationals. After shrinking X around L suitably, we may assume that $K_X + \Delta$ is $\mathbb{R}$-Cartier for every $\Delta \in \mathcal{R}(V;L)$ since V is finite-dimensional and L is compact. Let Θ be the union of the support of any element of $\mathcal{R}(V;L)$. By [BieM2, Theorem 13.2], after shrinking X around L suitably, we can construct a projective bimeromorphic morphism $f\colon Y \to X$ from a smooth complex analytic space Y such that $\mathrm{Exc}(f)$ and $\mathrm{Exc}(f) \cup \mathrm{Supp}\, f_*^{-1}\Theta$ are simple normal crossing divisors on Y, where $\mathrm{Exc}(f)$ denotes the exceptional locus of $f\colon Y \to X$. Thus, for any $\Delta \in \mathcal{R}(V;L)$, we can write

$$K_Y + \Delta_Y := f^*(K_X + \Delta)$$

such that $\mathrm{Supp}\,\Delta_Y$ is a simple normal crossing divisor on Y. In this situation, (X, Δ) is log canonical at L if and only if Δ is effective at L and the coefficients of Δ_Y are less than or equal to one over L. Hence, we can easily check that

$$\mathcal{L}(V;L) := \{\Delta \in V \mid K_X + \Delta \text{ is log canonical at } L\}$$

is a rational polytope contained in $\mathcal{R}(V;L)$. We can also check that there exists an open neighborhood U of L such that $(U, \Delta|_U)$ is log canonical for every $\Delta \in \mathcal{L}(V;L)$.

2.1.11. Let X be a complex analytic space. An *analytic subset* (resp. A *locally closed analytic subset*) of X is the support of a closed analytic subspace (resp. a locally closed analytic subspace) of X. A *Zariski open subset* of X means the complement of an analytic subset. We note the following easy example.

Example 2.1.12. We consider $\Delta := \{z \in \mathbb{C} \mid |z| < 1\}$ and $\Delta^* := \Delta \setminus \{0\}$. Then Δ^* is a Zariski open subset of Δ. We put

$$U := \Delta^* \setminus \left\{ \frac{1}{n} \,\middle|\, n \in \mathbb{Z} \text{ with } n \geq 2 \right\}.$$

Then U is a Zariski open subset of Δ^* since

$$\left\{ \frac{1}{n} \,\middle|\, n \in \mathbb{Z} \text{ with } n \geq 2 \right\}$$

is a closed analytic subset of Δ^*. However, U is not a Zariski open subset of Δ. This is because

$$\{0\} \cup \left\{ \frac{1}{n} \,\middle|\, n \in \mathbb{Z} \text{ with } n \geq 2 \right\}$$

is not a closed analytic subset of Δ.

2.1.13. A subset $\mathcal{S}$ of a complex analytic space X is said to be *analytically meagre* if

$$\mathcal{S} \subset \bigcup_{n \in \mathbb{N}} Y_n,$$

where each Y_n is a locally closed analytic subset of X of codimension ≥ 1.

Let X be a complex analytic space. We say that a property P holds for an analytically sufficiently general point $x \in X$ when P holds for every point x contained in $X \setminus \mathcal{S}$ for some analytically meagre subset $\mathcal{S}$ of X.

Let $\pi\colon X \to Y$ be a morphism of analytic spaces. Similarly, we say that a property P holds for an *analytically sufficiently general fiber* of $\pi\colon X \to Y$ when P holds for $\pi^{-1}(y)$ for every $y \in Y \setminus \mathcal{S}$, where $\mathcal{S}$ is some analytically meagre subset of Y.

In this book, we will freely use the following facts, which can be found in [BăS, Chapter III].

2.1.14. Let $\pi\colon X \to Y$ be a projective surjective morphism of complex analytic spaces and let $\mathcal{L}$ be a line bundle on X. If $R^p\pi_*\mathcal{L} = 0$ holds, then $H^p(F, \mathcal{L}|_F) = 0$ for an analytically sufficiently general fiber F of $\pi\colon X \to Y$. If $H^0(F, \mathcal{L}|_F) \neq 0$ for an analytically sufficiently general fiber F of $\pi\colon X \to Y$, then $\pi_*\mathcal{L} \neq 0$ holds.

We will use the following convention throughout this book.

2.1.15. The expression '... for every $m \gg 0$' means that 'there exists a positive real number m_0 such that ... for every $m \geq m_0$.'

2.2 Relatively ample and relatively nef ℝ-line bundles

In this section, we will collect some basic properties of relatively nef and relatively ample ℝ-line bundles for the reader's convenience. We will frequently use them in subsequent sections.

Let us recall the definition of *projective morphisms of complex analytic spaces* for the sake of completeness.

Definition 2.2.1 (Projective morphisms of complex analytic spaces)**.** Let $\pi\colon X \to Y$ be a proper morphism of complex analytic spaces and let $\mathcal{L}$ be a line bundle on X. Then $\mathcal{L}$ is said to be *π-very ample* or *relatively very ample over* Y if $\mathcal{L}$ is *π-free*, that is,

$$\pi^*\pi_*\mathcal{L} \to \mathcal{L}$$

is surjective, and the induced morphism

$$X \to \mathbb{P}_Y(\pi_*\mathcal{L})$$

over Y is a closed embedding. A line bundle $\mathcal{L}$ on X is called *π-ample* or *ample over Y* if for any point $y \in Y$ there are an open neighborhood U of y and a positive integer m such that $\mathcal{L}^{\otimes m}|_{\pi^{-1}(U)}$ is relatively very ample over U. Let D be a Cartier divisor on X. Then we say that D is *π-very ample*, *π-free*, and *π-ample* if the line bundle $\mathcal{O}_X(D)$ is so, respectively. We note that $\pi\colon X \to Y$ is said to be *projective* when there exists a π-ample line bundle on X.

For the basic properties of π-ample line bundles, see [BăS, Chapter IV] and [Na2, Chapter II. §1.c. Ample line bundles]. Since we are mainly interested in $\mathbb{R}$-line bundles in this book, the following easy lemma is indispensable.

Lemma 2.2.2. *Let $\pi\colon X \to Y$ be a projective morphism between complex analytic spaces and let W be a compact subset of Y. Let $\mathcal{L}$ be an $\mathbb{R}$-line bundle on X. Then the following two conditions are equivalent.*

(i) *$\mathcal{L}$ is π-ample over W, that is, $\mathcal{L}|_{\pi^{-1}(w)}$ is ample in the usual sense for every $w \in W$.*

(ii) *$\mathcal{L}$ is π-ample over some open neighborhood U of W, that is, $\mathcal{L}|_{\pi^{-1}(U)}$ is a finite positive $\mathbb{R}$-linear combination of $\pi|_{\pi^{-1}(U)}$-ample line bundles.*

Sketch of Proof of Lemma 2.2.2. It is obvious that (i) follows from (ii). Hence it is sufficient to prove that (ii) follows from (i). It is an easy exercise to modify the proof of [FM2, Lemmas 6.1 and 6.2] suitably with the aid of [Na1, Proposition 1.4]. □

Throughout this book, we will freely use Lemma 2.2.2 without mentioning it explicitly. The following lemma is more or less well known to the experts. We describe it here for the sake of completeness.

Lemma 2.2.3. *Let $\pi\colon X \to Y$ be a projective surjective morphism of complex analytic spaces such that X and Y are both irreducible. Let $\mathcal{L}$ be a line bundle on X. Assume that $\mathcal{L}|_{\pi^{-1}(y)}$ is ample for some $y \in Y$. Then there exists a Zariski open neighborhood U of y in Y and a positive integer m such that $\mathcal{L}^{\otimes m}|_{\pi^{-1}(U)}$ is π-very ample over U. In particular, $\mathcal{L}|_{\pi^{-1}(U)}$ is π-ample over U.*

Proof. It is well known that there exists a small open neighborhood U_1 of y in Y such that $\mathcal{L}|_{\pi^{-1}(U_1)}$ is π-ample over U_1 (see [Na1, Proposition 1.4]). Therefore, we can take some positive integer m such that $\mathcal{L}^{\otimes m}$ is π-very ample over some small open neighborhood U_2 of y in Y. We consider $\pi^*\pi_*\mathcal{L}^{\otimes m} \to \mathcal{L}^{\otimes m}$. It is obviously surjective over U_2. Therefore,

$$\pi\left(\operatorname{Supp}\operatorname{Coker}(\pi^*\pi_*\mathcal{L}^{\otimes m} \to \mathcal{L}^{\otimes m})\right) \cap U_2 = \emptyset.$$

Then we put

$$U_3 := Y \setminus \pi\left(\operatorname{Supp}\operatorname{Coker}(\pi^*\pi_*\mathcal{L}^{\otimes m} \to \mathcal{L}^{\otimes m})\right).$$

Hence U_3 is a non-empty Zariski open subset of Y such that $y \in U_3$ and that $\pi^*\pi_*\mathcal{L}^{\otimes m} \to \mathcal{L}^{\otimes m}$ is surjective over U_3. We put

$$\mathcal{I} := \operatorname{Im}\left(\pi^*\pi_*\mathcal{L}^{\otimes m} \to \mathcal{L}^{\otimes m}\right) \otimes \mathcal{L}^{\otimes(-m)} \subset \mathcal{O}_X.$$

Then $\mathcal{I}$ is a coherent ideal sheaf on X. We take the blow-up $p\colon Z \to X$ of X along the ideal sheaf $\mathcal{I}$, that is, $p\colon Z := \mathrm{Projan}_X \bigoplus_{d=0}^{\infty} \mathcal{I}^d \to X$. By construction,

$$\mathcal{M} := \mathrm{Im}\left(p^*\pi^*\pi_*\mathcal{L}^{\otimes m} \to p^*\mathcal{L}^{\otimes m}\right)$$

becomes a line bundle on Z. This gives a closed embedding

$$Z \simeq \mathbb{P}_Z(\mathcal{M}) \hookrightarrow \mathbb{P}_Y(\pi_*\mathcal{L}^{\otimes m}) \times_Y Z.$$

Thus we obtain a morphism $\alpha\colon Z \to \mathbb{P}_Y(\pi_*\mathcal{L}^{\otimes m})$ over Y. By construction again, p is an isomorphism over U_3 and α is a closed embedding over U_2. We can take a non-empty Zariski open subset V of $\alpha(Z)$ such that α is flat over V. Without loss of generality, we may assume that V contains $q^{-1}(U_2)$, where $q\colon \alpha(Z) \to Y$, and that α is an isomorphism over V.

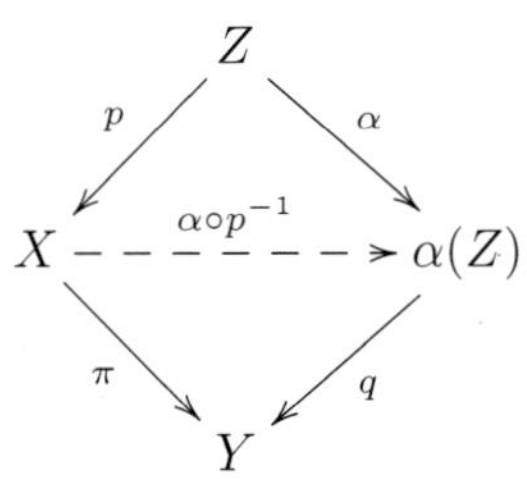

We put

$$U := U_3 \cap (Y \setminus q\left(\alpha(Z) \setminus V\right)).$$

Then U is a non-empty Zariski open subset of Y such that $y \in U$ and $\alpha \circ p^{-1}\colon X \dashrightarrow \mathbb{P}_Y(\pi_*\mathcal{L}^{\otimes m})$ is a closed embedding over U. Therefore, $\mathcal{L}^{\otimes m}$ is π-very ample over U. $\square$

As an application of Lemma 2.2.3, we have:

Lemma 2.2.4. *Let $\pi\colon X \to Y$ be a projective surjective morphism of complex analytic spaces. Let $\mathcal{L}$ be an $\mathbb{R}$-line bundle on X. Assume that $\mathcal{L}|_{\pi^{-1}(y)}$ is ample for some $y \in Y$. Then there exists a Zariski open neighborhood U of y in Y such that $\mathcal{L}|_{\pi^{-1}(U)}$ is π-ample over U.*

In the theory of minimal models, we have to treat $\mathbb{R}$-line bundles. Therefore, Lemma 2.2.4 is indispensable. Since we can not directly apply geometric arguments to $\mathbb{R}$-line bundles, Lemma 2.2.4 is not so obvious.

Proof of Lemma 2.2.4. We can write $\mathcal{L} = \sum_{i\in I} a_i\mathcal{L}_i$ in $\mathrm{Pic}(X) \otimes_{\mathbb{Z}} \mathbb{R}$ such that a_i is a positive real number, $\mathcal{L}_i \in \mathrm{Pic}(X)$, and $\mathcal{L}_i|_{\pi^{-1}(y)}$ is ample for every $i \in I$. Let $X = \bigcup_{j\in J} X_j$ be the irreducible decomposition. We put

$$J_1 := \{j \in J \mid y \in \pi(X_j)\} \quad \text{and} \quad J_2 := \{j \in J \mid y \notin \pi(X_j)\}.$$

We take an irreducible component X_j of X with $j \in J_1$. By applying Lemma 2.2.3 to $\mathcal{L}_i|_{X_j}$, we can find a Zariski closed subset Σ_j of $\pi(X_j)$ such that $y \in \pi(X_j) \setminus \Sigma_j$,

$\mathcal{L}_i|_{X_j}$ is ample over $\pi(X_j)\setminus\Sigma_j$ for every $i \in I$. This implies that $\mathcal{L}|_{X_j}$ is ample over $\pi(X_j)\setminus\Sigma_j$. We put

$$\Sigma := \left(\bigcup_{i\in J_1}\Sigma_j\right)\cup\pi\left(\bigcup_{j\in J_2}X_j\right).$$

Then Σ is a Zariski closed subset of Y such that $y \in Y\setminus\Sigma$ and that $\mathcal{L}$ is π-ample over $Y\setminus\Sigma$. Therefore, $U := Y\setminus\Sigma$ is a desired Zariski open neighborhood of y in Y. □

By Lemma 2.2.4, we can easily obtain:

Lemma 2.2.5. *Let $\pi\colon X \to Y$ be a projective surjective morphism of complex analytic spaces. Let $\mathcal{L}$ be an $\mathbb{R}$-line bundle on X. Assume that $\mathcal{L}|_{\pi^{-1}(y_0)}$ is nef for some $y_0 \in Y$. Then there exists an analytically meagre subset $\mathcal{S}$ such that $\mathcal{L}|_{\pi^{-1}(y)}$ is nef for every $y \in Y\setminus\mathcal{S}$.*

Although Lemma 2.2.5 is easy, it will play a very important role in our framework of the minimal model program of complex analytic spaces. We note that we can not make $\mathcal{S}$ a Zariski closed subset of Y in Lemma 2.2.5.

Proof of Lemma 2.2.5. We take a π-ample line bundle $\mathcal{H}$ on X. Then $(m\mathcal{L}+\mathcal{H})|_{\pi^{-1}(y_0)}$ is ample for every positive integer m. Therefore, by Lemma 2.2.4, for each $m \in \mathbb{Z}_{>0}$, we can take a Zariski open neighborhood U_m of y_0 in Y such that $m\mathcal{L}+\mathcal{H}$ is π-ample over U_m. We put $\mathcal{S} := \bigcup_{m\in\mathbb{Z}_{>0}}(Y\setminus U_m)$. Then $(m\mathcal{L}+\mathcal{H})|_{\pi^{-1}(y)}$ is ample for every $m \in \mathbb{Z}_{>0}$ and every $y \in Y\setminus\mathcal{S}$. This means that $\mathcal{L}|_{\pi^{-1}(y)}$ is nef for every $y \in Y\setminus\mathcal{S}$. □

The following obvious corollary of Lemma 2.2.5 is also useful for geometric applications. We note that we sometimes have to treat a countably infinite set of line bundles.

Corollary 2.2.6. *Let $\pi\colon X \to Y$ be a projective surjective morphism of complex analytic spaces. Let $\mathcal{L}_i$ be an $\mathbb{R}$-line bundle on X for $i \in \mathbb{N}$. Assume that $\mathcal{L}_i|_{\pi^{-1}(y_0)}$ is nef for some $y_0 \in Y$ and for every $i \in \mathbb{N}$. Then there exists an analytically meagre subset $\mathcal{S}$ such that $\mathcal{L}_i|_{\pi^{-1}(y)}$ is nef for every $y \in Y\setminus\mathcal{S}$ and every $i \in \mathbb{N}$. Therefore, if $\mathcal{H}$ is any π-ample $\mathbb{R}$-line bundle on X, then $(\mathcal{H}+\mathcal{L}_i)|_{\pi^{-1}(y)}$ is ample for every $y \in Y\setminus\mathcal{S}$ and every $i \in \mathbb{N}$.*

Proof. By Lemma 2.2.5, for each $i \in \mathbb{N}$, we can find an analytically meagre subset $\mathcal{S}_i$ of Y such that $\mathcal{L}_i|_{\pi^{-1}(y)}$ is nef for every $y \in Y\setminus\mathcal{S}_i$. We put $\mathcal{S} := \bigcup_{i\in\mathbb{N}}\mathcal{S}_i$. Then it is easy to see that $\mathcal{S}$ is a desired analytically meagre subset of Y. □

By the proof of Lemma 2.2.5 and Corollary 2.2.6, we have:

Remark 2.2.7. In Lemma 2.2.5 and Corollary 2.2.6, we can make $Y\setminus\mathcal{S}$ a countable intersection of non-empty Zariski open subsets of Y.

2.3 Non-lc ideal sheaves

Let us recall the notion of *non-lc ideal sheaves*. It is well defined even in the complex analytic setting.

Definition 2.3.1 (Non-lc ideal sheaves, see [Fu4, Definition 7.1]). Let X be a normal complex variety and let Δ be an effective $\mathbb{R}$-divisor on X such that $K_X+\Delta$ is $\mathbb{R}$-Cartier. Let $f\colon Z \to X$ be a projective bimeromorphic morphism from a smooth complex variety Z with $K_Z+\Delta_Z := f^*(K_X+\Delta)$ such that $\operatorname{Supp}\Delta_Z$ is a simple normal crossing divisor on Z. Then we put

$$\mathcal{J}_{\mathrm{NLC}}(X,\Delta) := f_*\mathcal{O}_Z(\lceil -(\Delta_Z^{<1})\rceil - \lfloor \Delta_Z^{>1}\rfloor) = f_*\mathcal{O}_Z(-\lfloor \Delta_Z\rfloor + \Delta_Z^{=1})$$

and call it the *non-lc ideal sheaf associated to* (X,Δ). We put

$$\mathcal{J}(X,\Delta) := f_*\mathcal{O}_Z(-\lfloor \Delta_Z\rfloor).$$

Then $\mathcal{J}(X,\Delta)$ is the well-known *multiplier ideal sheaf associated to* (X,Δ). By definition, the following inclusion

$$\mathcal{J}(X,\Delta) \subset \mathcal{J}_{\mathrm{NLC}}(X,\Delta)$$

always holds. By definition again, we can easily see that the support of

$$\mathcal{O}_X/\mathcal{J}_{\mathrm{NLC}}(X,\Delta)$$

is the non-lc locus $\mathrm{Nlc}(X,\Delta)$ of (X,Δ).

By the standard argument (see, for example, [Fu4, Lemma 7.2]), there are no difficulties to check the following lemma.

Lemma 2.3.2. *In Definition 2.3.1, $\mathcal{J}_{\mathrm{NLC}}(X,\Delta)$ and $\mathcal{J}(X,\Delta)$ are independent of the choice of the resolution $f\colon Z \to X$. Hence $\mathcal{J}_{\mathrm{NLC}}(X,\Delta)$ and $\mathcal{J}(X,\Delta)$ are well-defined coherent ideal sheaves on X.*

Sketch of Proof of Lemma 2.3.2. Since we do not use $\mathcal{J}(X,\Delta)$ in this book and the proof for $\mathcal{J}(X,\Delta)$ is simpler than for $\mathcal{J}_{\mathrm{NLC}}(Z,\Delta)$, we only treat $\mathcal{J}_{\mathrm{NLC}}(X,\Delta)$ here. Let $f_1\colon Z_1 \to X$ and $f_2\colon Z_2 \to X$ be two resolutions with $K_{Z_1}+\Delta_{Z_1} = f_1^*(K_X+\Delta)$ and $K_{Z_2}+\Delta_{Z_2} = f_2^*(K_X+\Delta)$ as in Definition 2.3.1. We take an arbitrary point $x \in X$. It is sufficient to prove that

$$f_{1*}\mathcal{O}_{Z_1}(-\lfloor \Delta_{Z_1}\rfloor + \Delta_{Z_1}^{=1}) = f_{2*}\mathcal{O}_{Z_2}(-\lfloor \Delta_{Z_2}\rfloor + \Delta_{Z_2}^{=1})$$

holds on some open neighborhood of x. Therefore, by shrinking X around x and taking an elimination of indeterminacy of $Z_2 \dashrightarrow Z_1$, we may further assume that f_2 decomposes as

$$f_2\colon Z_2 \longrightarrow Z_1 \xrightarrow{f_1} X.$$

Then, by [Fu6, Proposition 6.3.1], we can directly check that

$$f_{1*}\mathcal{O}_{Z_1}(-\lfloor \Delta_{Z_1}\rfloor + \Delta_{Z_1}^{=1}) = f_{2*}\mathcal{O}_{Z_2}(-\lfloor \Delta_{Z_2}\rfloor + \Delta_{Z_2}^{=1})$$

holds. We finish the proof. □

In this book, we need the following Bertini-type theorem for $\mathcal{J}_{\mathrm{NLC}}(X, \Delta)$.

Lemma 2.3.3 ([Fu4, Proposition 7.5]). *Let X be a normal complex variety and let Δ be an effective $\mathbb{R}$-divisor on X such that $K_X + \Delta$ is $\mathbb{R}$-Cartier. Let Λ ($\simeq \mathbb{P}^N$) be a finite-dimensional linear system on X. Let $X^\dagger$ be any relatively compact open subset of X. Then there exists an analytically meagre subset $\mathcal{S}$ of Λ such that*

$$\mathcal{J}_{\mathrm{NLC}}(X^\dagger, \Delta + tD) = \mathcal{J}_{\mathrm{NLC}}(X^\dagger, \Delta)$$

holds outside the base locus $\mathrm{Bs}\,\Lambda$ of Λ for every element D of $\Lambda \setminus \mathcal{S}$ and every $0 \leq t \leq 1$.

Proof. Without loss of generality, we can freely replace X with a relatively compact open neighborhood of $\overline{X^\dagger}$. Therefore, by the desingularization theorem (see [BieM2, Theorem 13.2]), we can take a projective bimeromorphic morphism $f\colon Z \to X$ from a smooth complex variety Z with $K_Z + \Delta_Z = f^*(K_X + \Delta)$ such that $\mathrm{Supp}\,\Delta_Z$ is a simple normal crossing divisor on Z. By replacing X with $X \setminus \mathrm{Bs}\,\Lambda$, we may further assume that $\mathrm{Bs}\,\Lambda = \emptyset$. By Bertini's theorem, there exists an analytically meagre subset $\mathcal{S}$ of Λ such that f^*D is smooth, $f^*D = f^{-1}_*D$, f^*D and $\mathrm{Supp}\,\Delta_Z$ have no common irreducible components, and the support of $f^*D + \mathrm{Supp}\,\Delta_Z$ is a simple normal crossing divisor on Z for every element D of $\Lambda \setminus \mathcal{S}$. Then $K_Z + \Delta_Z + f^*tD = f^*(K_X + \Delta + tD)$ holds over $X^\dagger$ with $f^*tD = tf^{-1}_*D$. Thus,

$$\lceil -(\Delta_Z^{<1}) \rceil - \lfloor \Delta_Z^{>1} \rfloor = \lceil -(\Delta_Z + f^*tD)^{<1} \rceil - \lfloor (\Delta_Z + f^*tD)^{>1} \rfloor$$

holds over $X^\dagger$ for every $0 \leq t \leq 1$ and every element D of $\Lambda \setminus \mathcal{S}$. Thus, we obtain

$$\mathcal{J}_{\mathrm{NLC}}(X^\dagger, \Delta + tD) = \mathcal{J}_{\mathrm{NLC}}(X^\dagger, \Delta)$$

by definition. This is what we wanted. □

We give a small remark on Lemma 2.3.3.

Remark 2.3.4. Although the following expression is not so precise, we usually say that if D is a general member of Λ then $\mathcal{J}_{\mathrm{NLC}}(X^\dagger, \Delta + tD) = \mathcal{J}_{\mathrm{NLC}}(X^\dagger, \Delta)$ holds outside $\mathrm{Bs}\,\Lambda$ for simplicity. Hence, if $D_1, \ldots, D_k$ are general members of Λ and $0 \leq t_i \leq 1$ for every $1 \leq i \leq k$, then

$$\mathcal{J}_{\mathrm{NLC}}\left(X^\dagger, \Delta + \sum_{i=1}^{k} t_i D_i\right) = \mathcal{J}_{\mathrm{NLC}}(X^\dagger, \Delta)$$

holds outside $\mathrm{Bs}\,\Lambda$.

We need the following lemma in order to reduce the problems for $\mathbb{R}$-divisors to simpler problems for $\mathbb{Q}$-divisors.

Lemma 2.3.5. *Let X be a normal complex variety and let L be a compact subset of X. Let Δ be an effective $\mathbb{R}$-divisor on X such that $K_X + \Delta$ is $\mathbb{R}$-Cartier at L. Then, after shrinking X around L suitably, there exist effective $\mathbb{Q}$-divisors $\Delta_1, \ldots, \Delta_k$ on*

X and positive real numbers $r_1, \ldots, r_k$ with $\sum_{i=1}^k r_i = 1$ such that $K_X + \Delta_i$ is $\mathbb{Q}$-Cartier for every i, $\Delta = \sum_{i=1}^k r_i \Delta_i$, and $\mathcal{J}_{\mathrm{NLC}}(X, \Delta_i) = \mathcal{J}_{\mathrm{NLC}}(X, \Delta)$ holds for every i. In particular, if (X, Δ) is log canonical, then (X, Δ_i) is log canonical for every i.

Proof. By shrinking X around L suitably, we may assume that $\operatorname{Supp}\Delta$ has only finitely many irreducible components. Let $\operatorname{Supp}\Delta := \sum_{j=1}^l D_j$ be the irreducible decomposition. We consider the $\mathbb{R}$-vector space $V := \bigoplus_{j=1}^l \mathbb{R}D_j$. We put

$$\mathcal{R}(V; L) := \{D \in V \mid K_X + D \text{ is } \mathbb{R}\text{-Cartier at } L\}.$$

Then $\mathcal{R}(V; L)$ is an affine subspace of V defined over the rationals (see 2.1.10). By shrinking X around L suitably again, we may assume that $K_X + D$ is $\mathbb{R}$-Cartier for every $D \in \mathcal{R}(V; L)$. By [BieM2, Theorem 13.2], we may further assume that there exists a projective bimeromorphic morphism $f\colon Z \to X$ from a smooth complex analytic space Z such that $\operatorname{Exc}(f)$ and $\operatorname{Exc}(f) \cup \sum_{j=1}^l \operatorname{Supp} f_*^{-1} D_j$ are simple normal crossing divisors on Z. We put

$$\mathcal{S}_\Delta(V; L) := \left\{ D \in \mathcal{R}(V; L) \,\middle|\, \begin{array}{l} a(E, X, D) = a(E, X, \Delta) \text{ holds for every} \\ \text{divisor } E \text{ on } Z \text{ with } a(E, X, \Delta) \in \mathbb{Q} \end{array} \right\}.$$

Then $\mathcal{S}_\Delta(V; L)$ is an affine subspace of V defined over the rationals with $\Delta \in \mathcal{S}_\Delta(V; L)$. Since $\mathcal{S}_\Delta(V; L)$ is defined over the rationals, we can take effective $\mathbb{Q}$-divisors $\Delta_1, \ldots, \Delta_k$ from $\mathcal{S}_\Delta(V; L)$ such that they are close to Δ in $\mathcal{S}_\Delta(V; L)$ and positive real numbers $r_1, \ldots, r_k$ with all the desired properties. □

Although we do not use multiplier ideal sheaves in this book, we note:

Remark 2.3.6. In Lemma 2.3.5, we see that $\mathcal{J}(X, \Delta_i) = \mathcal{J}(X, \Delta)$ holds for every i by construction. In particular, (X, Δ_i) is kawamata log terminal for every i if (X, Δ) is kawamata log terminal.

Chapter 3
Vanishing theorems

In this chapter, we will treat vanishing theorems. In Section 3.1, we will briefly review some vanishing theorems for analytic simple normal crossing pairs. Theorem 3.1.5 is one of the main ingredients of this book. The reader can find all the details in [Fu15] (see also [FF] and [Fu18]). For a completely different approach due to Murayama, see [Mu]. In Section 3.2, we will explain vanishing theorems of Reid–Fukuda type although we do not need them in this book. Then, in Section 3.3, we will establish some useful vanishing theorems for normal pairs by using the results explained in Section 3.1. As an application of Section 3.3, we will prove basic properties of log canonical centers in Section 3.4. They will play a crucial role in this book.

3.1 Vanishing theorems for analytic simple normal crossing pairs

In this section, let us quickly recall some vanishing theorems established in [Fu15] (see also [FF] and [Fu18]). We start with the definition of *analytic simple normal crossing pairs*.

Definition 3.1.1 (Analytic simple normal crossing pairs)**.** Let X be a simple normal crossing divisor on a smooth complex analytic space M and let B be an $\mathbb{R}$-divisor on M such that $\operatorname{Supp}(B+X)$ is a simple normal crossing divisor on M and that B and X have no common irreducible components. Then we put $D := B|_X$ and consider the pair (X, D). We call (X, D) an *analytic globally embedded simple normal crossing pair* and M the *ambient space* of (X, D).

If the pair (X, D) is locally isomorphic to an analytic globally embedded simple normal crossing pair at any point of X and the irreducible components of X and D are all smooth, then (X, D) is called an *analytic simple normal crossing pair*.

When (X, D) is an analytic simple normal crossing pair, X has an invertible dualizing sheaf ω_X since it is Gorenstein. We use the symbol K_X as a formal divisor class with an isomorphism $\mathcal{O}_X(K_X) \simeq \omega_X$ if there is no danger of confusion. Note that we can not always define K_X globally with $\mathcal{O}_X(K_X) \simeq \omega_X$. In general, it only exists locally on X.

Remark 3.1.2. Let X be a smooth complex analytic space and let D be an $\mathbb{R}$-divisor on X such that $\operatorname{Supp} D$ is a simple normal crossing divisor on X. Then, by considering $M := X \times \mathbb{C}$, we can see (X, D) as an analytic globally embedded simple normal crossing pair.

The notion of *strata*, which is a generalization of that of log canonical centers, plays a crucial role.

Definition 3.1.3 (Strata)**.** Let (X, D) be an analytic simple normal crossing pair such that D is effective. Let $\nu\colon X^\nu \to X$ be the normalization. We put

$$K_{X^\nu} + \Theta = \nu^*(K_X + D).$$

This means that Θ is the union of $\nu_*^{-1}D$ and the inverse image of the singular locus of X. If S is an irreducible component of X or the ν-image of some log canonical center of (X^ν, Θ), then S is called a *stratum* of (X, D).

We recall Siu's theorem on coherent analytic sheaves, which is a special case of [Si, Theorem 4].

Theorem 3.1.4. *Let $\mathcal{F}$ be a coherent sheaf on a complex analytic space X. Then there exists a locally finite family $\{Y_i\}_{i\in I}$ of complex analytic subvarieties of X such that*

$$\mathrm{Ass}_{\mathcal{O}_{X,x}}(\mathcal{F}_x) = \{\mathfrak{p}_{x,1}, \ldots, \mathfrak{p}_{x,r(x)}\}$$

holds for every point $x \in X$, where $\mathfrak{p}_{x,1}, \ldots, \mathfrak{p}_{x,r(x)}$ are the prime ideals of $\mathcal{O}_{X,x}$ associated to the irreducible components of the germs $Y_{i,x}$ of Y_i at x with $x \in Y_i$. We note that each Y_i is called an associated subvariety *of $\mathcal{F}$.*

Now we are ready to state the main result of [Fu15].

Theorem 3.1.5 ([Fu15, Theorem 1.1])**.** *Let (X, Δ) be an analytic simple normal crossing pair such that Δ is a boundary $\mathbb{R}$-divisor on X. Let $f\colon X \to Y$ be a projective morphism to a complex analytic space Y and let $\mathcal{L}$ be a line bundle on X. Let q be an arbitrary non-negative integer. Then we have the following properties.*

(1) (*Strict support condition*). *If $\mathcal{L} - (\omega_X + \Delta)$ is f-semiample, then every associated subvariety of $R^q f_*\mathcal{L}$ is the f-image of some stratum of (X, Δ).*
(2) (*Vanishing theorem*). *If $\mathcal{L} - (\omega_X + \Delta) \sim_{\mathbb{R}} f^*\mathcal{H}$ holds for some π-ample $\mathbb{R}$-line bundle $\mathcal{H}$ on Y, where $\pi\colon Y \to Z$ is a projective morphism to a complex analytic space Z, then we have*

$$R^p\pi_* R^q f_*\mathcal{L} = 0$$

for every $p > 0$.

We note that Theorem 3.1.5 is one of the main ingredients of this book. Or, we can see this book as an application of Theorem 3.1.5. We make a supplementary remark on Theorem 3.1.5.

Remark 3.1.6. In Theorem 3.1.5 (and Theorem 3.2.3 below), we always assume that Δ is *globally* $\mathbb{R}$*-Cartier*, that is, Δ is a finite $\mathbb{R}$-linear combination of Cartier divisors. We note that if the support of Δ has only finitely many irreducible components then it is globally $\mathbb{R}$-Cartier. Since we are mainly interested in the standard setting explained in 1.1.3, this assumption is harmless to geometric applications. Under this assumption, we can obtain an $\mathbb{R}$-line bundle $\mathcal{N}$ on X naturally associated to

$\mathcal{L} - (\omega_X + \Delta)$. The assumption in (1) means that $\mathcal{N}$ is a finite positive $\mathbb{R}$-linear combination of π-semiample line bundles on X. The assumption in (2) says that $\mathcal{N} = f^*\mathcal{H}$ holds in $\mathrm{Pic}(X) \otimes_{\mathbb{Z}} \mathbb{R}$.

A key result for the proof of Theorem 3.1.5 is Takegoshi's analytic generalization of Kollár's torsion-free and vanishing theorem (see [Ta]). For an alternative approach to Theorem 3.1.7, see [Fu5, Corollaries 1.2 and 1.5].

Theorem 3.1.7 (see [Ta]). *Let $f\colon X \to Y$ be a projective surjective morphism from a smooth irreducible complex analytic space X. Then $R^q f_*\omega_X$ is a torsion-free coherent sheaf on Y for every q.*

Furthermore, let $\pi\colon Y \to Z$ be a projective morphism between complex analytic spaces and let $\mathcal{A}$ be a π-ample line bundle on Y. Then

$$R^p\pi_*\left(\mathcal{A} \otimes R^q f_*\omega_X\right) = 0$$

holds for every $p > 0$ and every q.

If Y is projective and Z is a point, then Theorem 3.1.7 is nothing but Kollár's famous torsion-free and vanishing theorem. Note that Takegoshi established much more general results than Theorem 3.1.7 in [Ta] and the arguments in [Ta] are analytic. An alternative approach to Theorem 3.1.7 in [Fu5] is also analytic. In [Fu15], we first proved the following generalization of Theorem 3.1.7 with the aid of Saito's theory of mixed Hodge modules (see [Sa1], [Sa2], [Sa3], [FFS], and [Sa4]).

Theorem 3.1.8. *Let (X, D) be an analytic simple normal crossing pair such that D is reduced and let $f\colon X \to Y$ be a projective morphism of complex analytic spaces. Then we have the following properties.*

(1) (*Strict support condition*). *Every associated subvariety of $R^q f_*\omega_X(D)$ is the f-image of some stratum of (X, D) for every q.*
(2) (*Vanishing theorem*). *Let $\pi\colon Y \to Z$ be a projective morphism between complex analytic spaces and let $\mathcal{A}$ be a π-ample line bundle on Y. Then*

$$R^p\pi_*\left(\mathcal{A} \otimes R^q f_*\omega_X(D)\right) = 0$$

holds for every $p > 0$ and every q.

From now on, let us explain how to prove Theorem 3.1.8 from Theorem 3.1.7 with the aid of Saito's theory of mixed Hodge modules. In order to prove Theorem 3.1.8, we need the following description of strata.

3.1.9. Let (X, D) be an analytic simple normal crossing pair such that D is reduced. For any positive integer k, we put

$$X^{[k]} := \{x \in X \,|\, \mathrm{mult}_x X \geq k\}^{\sim},$$

where $Z^{\sim}$ denotes the normalization of Z. Then $X^{[k]}$ is the disjoint union of the intersections of k irreducible components of X, and is smooth. We have a reduced simple normal crossing divisor $D^{[k]} \subset X^{[k]}$ defined by the pull-back of D by the natural morphism $X^{[k]} \to X$. For any non-negative integer l, we put

$$D^{[k,l]} := \left\{ x \in X^{[k]} \,\middle|\, \operatorname{mult}_x D^{[k]} \geq l \right\}^{\sim}.$$

We note that $D^{[k,0]} = X^{[k]}$ holds by definition. We also note that $\dim D^{[k,l]} = n + 1 - k - l$, where $n = \dim X$. In this situation, S is a stratum of (X, D) if and only if S is the image of an irreducible component of $D^{[k,l]}$ for some $k > 0$ and $l \geq 0$.

As an application of Saito's theory of mixed Hodge modules, we established the following theorem in [FFS].

Theorem 3.1.10 ([FFS, Corollary 1 and 4.7. Remark]). *Let (X, D) be an analytic simple normal crossing pair with $\dim X = n$ such that D is reduced and let $f\colon X \to Y$ be a projective morphism to a smooth complex analytic space Y. Then there is the weight spectral sequence*

$$_FE_1^{-q,i+q} = \bigoplus_{k+l=n+q+1} R^i f_* \omega_{D^{[k,l]}/Y} \Rightarrow R^i f_* \omega_{X/Y}(D),$$

degenerating at E_2, and its E_1-differential d_1 splits so that the $_FE_2^{-q,i+q}$ are direct factors of $_FE_1^{-q,i+q}$.

Now we can easily check Theorem 3.1.8 by using Theorems 3.1.7 and 3.1.10.

Proof of Theorem 3.1.8. First we prove (1). Since the problem is local, we may assume that Y is a closed analytic subspace of a polydisc Δ^m. By replacing Y with Δ^m, we may further assume that Y itself is a polydisc. In this case, we can use Theorem 3.1.10. We note that $\omega_Y \simeq \mathcal{O}_Y$ holds. By Theorem 3.1.7,

$$_FE_1^{-q,i+q} \simeq \bigoplus_{k+l=n+q+1} R^i f_* \omega_{D^{[k,l]}}$$

satisfies the strict support condition, that is, every associated subvariety of

$$_FE_1^{-q,i+q} \simeq \bigoplus_{k+l=n+q+1} R^i f_* \omega_{D^{[k,l]}}$$

is the f-image of some stratum of (X, D). By Theorem 3.1.10, the associated subvariety of $_FE_2^{-q,i+q} = {_FE_\infty^{-q,i+q}}$ is the f-image of some stratum of (X, D). This implies that $R^q f_* \omega_X(D)$ satisfies the desired strict support condition. Next, we treat (2). We may assume that Z is a polydisc and Y is a closed analytic subspace of $Z \times \mathbb{P}^n$. By applying Theorem 3.1.10 to $f\colon X \to Y \hookrightarrow Z \times \mathbb{P}^n$, we obtain the following spectral sequence

$$E_1^{-q,i+q} = \bigoplus_{k+l=n+q+1} R^i f_* \omega_{D^{[k,l]}} \Rightarrow R^i f_* \omega_X(D)$$

which degenerates at E_2 such that its E_1-differential d_1 splits. By Theorem 3.1.7, we obtain

$$R^p\pi_*\left(\mathcal{A}\otimes E_1^{-q,i+q}\right)=0$$

for every $p>0$. Since the $E_2^{-q,i+q}=E_\infty^{-q,i+q}$ are direct factors of $E_1^{-q,i+q}$, we have

$$R^p\pi_*\left(\mathcal{A}\otimes E_2^{-q,i+q}\right)=0$$

for every $p>0$. This implies that

$$R^p\pi_*\left(\mathcal{A}\otimes R^qf_*\omega_X(D)\right)=0$$

holds for every $p>0$. This is what we wanted. □

The reader can find an alternative approach to Theorem 3.1.8 without using Saito's theory of mixed Hodge modules in [FF]. For a completely different approach due to Murayama, see [Mu]. Once we know Theorem 3.1.8, it is not difficult to prove Theorem 3.1.5 for the case where X is irreducible. Here we give a proof of Theorem 3.1.5 under the assumption that X is irreducible for the reader's convenience.

Proof of Theorem 3.1.5 when X is irreducible. We first prove (1). We take an arbitrary point $y\in Y$. It is sufficient to prove the strict support condition on a small Stein open neighborhood of y. Therefore, we will freely shrink Y around y suitably without mentioning it explicitly throughout this proof. By replacing Y with a small relatively compact Stein open neighborhood of y, we may assume that there exist Cartier divisors L and K_X such that $\mathcal{O}_X(L)\simeq\mathcal{L}$ and $\mathcal{O}_X(K_X)\simeq\omega_X$, respectively. By perturbing the coefficients of Δ suitably, we may further assume that Δ is a $\mathbb{Q}$-divisor. We put $N:=L-(K_X+\Delta)$. Then N is an f-semiample $\mathbb{Q}$-Cartier $\mathbb{Q}$-divisor on X such that $L=K_X+\Sigma+\lceil N\rceil$, where $\Sigma:=\lfloor\Delta\rfloor$, with $\{N\}=\{-\Delta\}$. By Kawamata's covering trick, we can take a finite cover $p\colon V\to X$ such that V is smooth, $\Sigma_V:=p^*\Sigma$ is a reduced simple normal crossing divisor on V, $N_V:=p^*N$ is Cartier, and $\mathcal{O}_X(L)=\mathcal{O}_X(K_X+\Sigma+\lceil N\rceil)$ is a direct summand of $p_*\mathcal{O}_V(K_V+\Sigma_V+N_V)$. Since N_V is an $(f\circ p)$-semiample Cartier divisor on V, we can take a finite cyclic cover $q\colon W\to V$ such that W is smooth, $\Sigma_W:=q^*\Sigma_V=q^*p^*\Sigma$ is a reduced simple normal crossing divisor on W, and $\mathcal{O}_V(K_V+\Sigma_V+N_V)$ is a direct summand of $q_*\mathcal{O}_W(K_W+\Sigma_W)$. By Theorem 3.1.8 (1), every associated subvariety of $R^q(f\circ p\circ q)_*\mathcal{O}_W(K_W+\Sigma_W)$ is the $(f\circ p\circ q)$-image of some stratum of (W,Σ_W). This implies that every associated subvariety of $R^qf_*\mathcal{O}_X(L)$ is the f-image of some stratum of (X,Δ). This is what we wanted.

From now, we will prove (2). As in the above proof of (1), we take an arbitrary point $z\in Z$ and consider a small relatively compact Stein open neighborhood of z. Throughout this proof, we will freely shrink Z around z without mentioning it explicitly. By perturbing the coefficients, we may assume that H is a π-ample $\mathbb{Q}$-divisor on Y with $L-(K_X+\Delta)\sim_{\mathbb{Q}}f^*H$, where $\mathcal{L}\simeq\mathcal{O}_X(L)$ and $\omega_X\simeq\mathcal{O}_X(K_X)$. We put $N:=L-(K_X+\Delta)$ and $\Sigma:=\lfloor\Delta\rfloor$. By Kawamata's covering trick, we can construct a finite cover $p\colon V\to X$ such that V is smooth, $\Sigma_V:=p^*\Sigma$ is a simple normal crossing divisor, and $N_V := p^*N$ is a Cartier divisor on V. Since $\mathcal{O}_X(L) = \mathcal{O}_X(K_X + \Sigma + \lceil N \rceil)$ is a direct summand of $p_*\mathcal{O}_V(K_V + \Sigma_V + N_V)$, it is sufficient to prove that

$$R^p\pi_*R^q(f \circ p)_*\mathcal{O}_V(K_V + \Sigma_V + N_V) = 0$$

for every $p > 0$. As is well known, we can take a finite composite of cyclic covers $q\colon W \to V$ such that W is smooth, $N_W := q^*N_V$ is $(\pi \circ f \circ p \circ q)$-free, $\Sigma_W := q^*\Sigma_V$ is a reduced simple normal crossing divisor, and $\mathcal{O}_V(K_V + \Sigma_V + N_V)$ is a direct summand of $q_*\mathcal{O}_W(K_W + \Sigma_W + N_W)$. Hence it is sufficient to prove that

$$R^p\pi_*R^q(f \circ p \circ q)_*\mathcal{O}_W(K_W + \Sigma_W + N_W) = 0$$

for every $p > 0$. We take the contraction morphism over Z associated to the surjection

$$(\pi \circ f \circ p \circ q)^*(\pi \circ f \circ p \circ q)_*\mathcal{O}_W(N_W) \to \mathcal{O}_W(N_W)$$

and take its Stein factorization. Then we have the following commutative diagram:

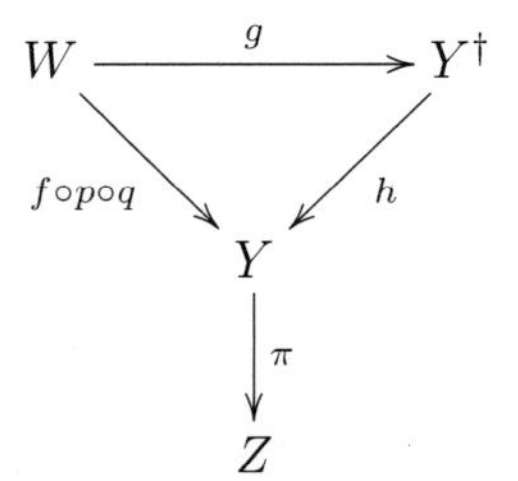

such that $g_*\mathcal{O}_W \simeq \mathcal{O}_{Y^\dagger}$, $\mathcal{O}_W(N_W) \simeq g^*\mathcal{A}$, where $\mathcal{A}$ is a $(\pi \circ h)$-ample line bundle on $Y^\dagger$, and h is finite. Then we obtain

$$\begin{aligned} & R^p\pi_*R^q(f \circ p \circ q)_*\mathcal{O}_W(K_W + \Sigma_W + N_W) \\ & \simeq R^p\pi_*R^q(h \circ g)_*\mathcal{O}_W(K_W + \Sigma_W + N_W) \\ & \simeq R^p\pi_*\left(h_*R^qg_*\mathcal{O}_W(K_W + \Sigma_W + N_W)\right) \\ & \simeq R^p(\pi \circ h)_*\left(\mathcal{A} \otimes R^qg_*\mathcal{O}_W(K_W + \Sigma_W)\right) \\ & = 0 \end{aligned}$$

by Theorem 3.1.8 (2). We finish the proof. □

In this book, we do not prove Theorem 3.1.5 for the case where X is reducible. For the details of the proof of Theorem 3.1.5, see [Fu15]. We will see that Theorem 3.1.5 for *varieties* is almost sufficient in this book.

3.2 Vanishing theorems of Reid–Fukuda type

Although we do not need vanishing theorems of Reid–Fukuda type in this book, we will shortly discuss them here for future usage. Let us start with the definition of *big* $\mathbb{R}$-line bundles for the reader's convenience.

Definition 3.2.1 (Big $\mathbb{R}$-line bundles)**.** Let $\pi\colon X \to Y$ be a projective morphism of complex analytic spaces such that X is irreducible and let $\mathcal{L}$ be an $\mathbb{R}$-line bundle on X. If $\mathcal{L}$ is a finite positive $\mathbb{R}$-linear combination of π-big line bundles on X, then $\mathcal{L}$ is said to be *big over* Y or *π-big over* Y.

We need the notion of *nef and log big $\mathbb{R}$-line bundles.*

Definition 3.2.2 (Nef and log big $\mathbb{R}$-line bundles)**.** Let (X, Δ) be an analytic simple normal crossing pair such that Δ is a boundary $\mathbb{R}$-divisor on X, let $f\colon X \to Y$ and $\pi\colon Y \to Z$ be projective morphisms between complex analytic spaces, and let $\mathcal{H}$ be an $\mathbb{R}$-line bundle on Y. Then $\mathcal{H}$ is said to be *nef and log big over Z with respect to $f\colon (X, \Delta) \to Y$* if $\mathcal{H} \cdot C \geq 0$ holds for every projective integral curve C on Y such that $\pi(C)$ is a point and $\mathcal{H}|_{f(S)}$ can be written as a finite positive $\mathbb{R}$-linear combination of π-big line bundles on $f(S)$ for every stratum S of (X, Δ).

We note that if $\mathcal{H}$ is π-ample in Definition 3.2.2 then it is obviously nef and log big over Z with respect to $f\colon (X, \Delta) \to Y$. Hence the following vanishing theorem is a generalization of Theorem 3.1.5 (2).

Theorem 3.2.3 (Vanishing theorem of Reid–Fukuda type, see [Fu15, Theorem 1.2])**.** *Let (X, Δ) be an analytic simple normal crossing pair such that Δ is a boundary $\mathbb{R}$-divisor on X. Let $f\colon X \to Y$ and $\pi\colon Y \to Z$ be projective morphisms between complex analytic spaces and let $\mathcal{L}$ be a line bundle on X. If $\mathcal{L} - (\omega_X + \Delta) \sim_{\mathbb{R}} f^*\mathcal{H}$ holds for some $\mathbb{R}$-line bundle $\mathcal{H}$ on Y which is nef and log big over Z with respect to $f\colon (X, \Delta) \to Y$, then*

$$R^p\pi_*R^qf_*\mathcal{L} = 0$$

holds for every $p > 0$ and every q.

The reader can find the detailed proof of Theorem 3.2.3 in [Fu15], which is harder than that of Theorem 3.1.5 (2). As an easy application of Theorem 3.2.3, we can establish the vanishing theorem of Reid–Fukuda type for log canonical pairs in the complex analytic setting.

Theorem 3.2.4 (Vanishing theorem of Reid–Fukuda type for log canonical pairs)**.** *Let (X, Δ) be a log canonical pair and let $\pi\colon X \to Y$ be a projective morphism of complex analytic spaces. Let L be a $\mathbb{Q}$-Cartier integral Weil divisor on X. Assume that $L - (K_X + \Delta)$ is nef and big over Y and that $(L - (K_X + \Delta))|_C$ is big over $\pi(C)$ for every log canonical center C of (X, Δ). Then*

$$R^q\pi_*\mathcal{O}_X(L) = 0$$

holds for every $q > 0$.

Proof. The proof of [Fu6, Theorem 5.7.6] works by Theorem 3.2.3. □

We leave the details of Theorems 3.2.3 and 3.2.4 for the interested readers. The vanishing theorems in the framework of quasi-log complex analytic spaces may be more useful (see [Fu16] and Theorems 6.1.4 and 6.1.5).

3.3 Vanishing theorems for normal pairs

In this section, we will prove some vanishing theorems, which are suitable for geometric applications. Theorem 3.3.1 will play a crucial role in subsequent sections. We note that the results in this section follow from Theorem 3.1.5.

Theorem 3.3.1 (see [Fu4, Theorem 8.1]). *Let $\pi\colon X \to Y$ be a projective morphism of complex analytic spaces such that X is a normal complex variety, and let W be a compact subset of Y. Let Δ be an effective $\mathbb{R}$-divisor on X such that $K_X + \Delta$ is $\mathbb{R}$-Cartier, and let $\mathcal{L}$ be a line bundle on X. We assume that $\mathcal{L} - (K_X + \Delta)$ is π-ample over W, that is, $(\mathcal{L} - (K_X + \Delta))|_{\pi^{-1}(w)}$ is ample for every $w \in W$. Let $\{C_i\}_{i\in I}$ be any set of log canonical centers of the pair (X, Δ). We put $V := \bigcup_{i\in I} C_i$ with the reduced structure. We further assume that V is disjoint from the non-lc locus $\mathrm{Nlc}(X, \Delta)$ of (X, Δ). Then there exists some open neighborhood U of W such that*

$$R^i\pi_*(\mathcal{J} \otimes \mathcal{L}) = 0$$

holds on U for every $i > 0$, where $\mathcal{J} := \mathcal{I}_V \cdot \mathcal{J}_{\mathrm{NLC}}(X, \Delta) \subset \mathcal{O}_X$ and $\mathcal{I}_V$ is the defining ideal sheaf of V on X. Therefore, the restriction map

$$\pi_*\mathcal{L} \to \pi_*(\mathcal{L}|_V) \oplus \pi_*(\mathcal{L}|_{\mathrm{Nlc}(X,\Delta)})$$

is surjective on U and

$$R^i\pi_*(\mathcal{L}|_V) = 0$$

holds on U for every $i > 0$. In particular, the restriction maps

$$\pi_*\mathcal{L} \to \pi_*(\mathcal{L}|_V)$$

and

$$\pi_*\mathcal{L} \to \pi_*(\mathcal{L}|_{\mathrm{Nlc}(X,\Delta)})$$

are surjective on U.

The result and argument in Step 1 in the proof of Theorem 3.3.1 is the most important part of this book. We will use it repeatedly in subsequent sections.

Proof of Theorem 3.3.1. In Steps 1 and 2, we will use the strict support condition (see Theorem 3.1.5 (1)) and the vanishing theorem (see Theorem 3.1.5 (2)), respectively. The assumption that $\mathcal{L} - (K_X + \Delta)$ is π-ample over W will be used only in Step 2.

We take an arbitrary point $w \in W$. Then it is sufficient to prove the desired vanishing theorem over some open neighborhood of w by the compactness of W. Therefore, we may replace Y with a relatively compact Stein open neighborhood of w and may assume that $\mathcal{L} - (K_X + \Delta)$ is π-ample over Y.

Step 1. We can take a resolution of singularities $f\colon Z \to X$ of X such that f is projective and that $\mathrm{Supp}\, f_*^{-1}\Delta \cup \mathrm{Exc}(f)$ is a simple normal crossing divisor on Z.

We may further assme that $f^{-1}(V)$ is a simple normal crossing divisor on Z. Then we can write

$$K_Z + \Delta_Z = f^*(K_X + \Delta).$$

Let T be the union of the irreducible components of $\Delta_Z^{=1}$ that are mapped into V by f. We consider the following short exact sequence

$$0 \to \mathcal{O}_Z(A - N - T) \to \mathcal{O}_Z(A - N) \to \mathcal{O}_T(A - N) \to 0,$$

where $A := \lceil -(\Delta_Z^{<1}) \rceil$ and $N := \lfloor \Delta_Z^{>1} \rfloor$. By definition, A is an effective f-exceptional divisor on Z. We obtain the following long exact sequence

$$\begin{aligned} 0 &\to f_*\mathcal{O}_Z(A - N - T) \to f_*\mathcal{O}_Z(A - N) \to f_*\mathcal{O}_T(A - N) \\ &\xrightarrow{\delta} R^1 f_*\mathcal{O}_Z(A - N - T) \to \cdots . \end{aligned}$$

Since

$$A - N - T - (K_Z + \{\Delta_Z\} + \Delta_Z^{=1} - T) = -(K_Z + \Delta_Z) \sim_{\mathbb{R}} -f^*(K_X + \Delta),$$

every associated subvariety of $R^1 f_*\mathcal{O}_Z(A - N - T)$ is the f-image of some stratum of $(Z, \{\Delta_Z\} + \Delta_Z^{=1} - T)$ by the strict support condition in Theorem 3.1.5 (1). Since $f^{-1}(V)$ is a simple normal crossing divisor, there are no strata of $(Z, \{\Delta_Z\} + \Delta_Z^{=1} - T)$ that are mapped into V by f. On the other hand, $V = f(T)$ holds by construction. Thus, the connecting homomorphism δ is a zero map. Hence we have a short exact sequence

$$0 \to f_*\mathcal{O}_Z(A - N - T) \to f_*\mathcal{O}_Z(A - N) \to f_*\mathcal{O}_T(A - N) \to 0. \tag{3.3.1}$$

We put $\mathcal{J} := f_*\mathcal{O}_Z(A - N - T) \subset \mathcal{O}_X$. Since V is disjoint from $\mathrm{Nlc}(X, \Delta)$ by assumption, the ideal sheaf $\mathcal{J}$ coincides with $\mathcal{I}_V$ and $\mathcal{J}_{\mathrm{NLC}}(X, \Delta)$ in a neighborhood of V and $\mathrm{Nlc}(X, \Delta)$, respectively. Therefore, we have $\mathcal{J} = \mathcal{I}_V \cdot \mathcal{J}_{\mathrm{NLC}}(X, \Delta)$. We note that if V is empty then $\mathcal{I}_V = \mathcal{O}_X$ and $\mathcal{J} = \mathcal{J}_{\mathrm{NLC}}(X, \Delta)$. We put $X^* := X \setminus \mathrm{Nlc}(X, \Delta)$ and $Z^* := f^{-1}(X^*)$. By restricting (3.3.1) to X^*, we obtain

$$0 \to f_*\mathcal{O}_{Z^*}(A - T) \to f_*\mathcal{O}_{Z^*}(A) \to f_*\mathcal{O}_T(A) \to 0.$$

Since $f_*\mathcal{O}_{Z^*}(A) \simeq \mathcal{O}_{X^*}$, we have $f_*\mathcal{O}_T(A) \simeq \mathcal{O}_V$. This implies that $\mathcal{O}_V \simeq f_*\mathcal{O}_T$ holds.

Step 2. Since

$$f^*\mathcal{L} + A - N - T - (K_Z + \{\Delta_Z\} + \Delta_Z^{=1} - T) \sim_{\mathbb{R}} f^*(\mathcal{L} - (K_X + \Delta)),$$

we have

$$R^i\pi_*(\mathcal{J} \otimes \mathcal{L}) \simeq R^i\pi_*(f_*\mathcal{O}_Z(A - N - T) \otimes \mathcal{L}) = 0$$

for every $i > 0$ by the vanishing theorem in Theorem 3.1.5 (2). If we put $V = \emptyset$, then we have $\mathcal{J} = \mathcal{J}_{\mathrm{NLC}}(X, \Delta)$. Therefore,

$$R^i\pi_*(\mathcal{J}_{\mathrm{NLC}}(X, \Delta) \otimes \mathcal{L}) = 0$$

holds for every $i > 0$ as a special case. By considering the short exact sequence

$$0 \to \mathcal{J} \to \mathcal{J}_{\mathrm{NLC}}(X, \Delta) \to \mathcal{O}_V \to 0,$$

we obtain

$$\cdots \to R^i\pi_*(\mathcal{J}_{\mathrm{NLC}}(X, \Delta) \otimes \mathcal{L}) \to R^i\pi_*(\mathcal{L}|_V) \to R^{i+1}\pi_*(\mathcal{J} \otimes \mathcal{L}) \to \cdots.$$

Since we have already checked

$$R^i\pi_*(\mathcal{J}_{\mathrm{NLC}}(X, \Delta) \otimes \mathcal{L}) = R^i\pi_*(\mathcal{J} \otimes \mathcal{L}) = 0$$

for every $i > 0$, we have $R^i\pi_*(\mathcal{L}|_V) = 0$ for all $i > 0$. Finally, we consider the following short exact sequence

$$0 \to \mathcal{J} \to \mathcal{O}_X \to \mathcal{O}_V \oplus \mathcal{O}_{\mathrm{Nlc}(X,\Delta)} \to 0.$$

By taking $\otimes \mathcal{L}$ and $R^i\pi_*$, we obtain that

$$0 \to \pi_*(\mathcal{J} \otimes \mathcal{L}) \to \pi_*\mathcal{L} \to \pi_*(\mathcal{L}|_V) \oplus \pi_*(\mathcal{L}|_{\mathrm{Nlc}(X,\Delta)}) \to 0$$

is exact.

We finish the proof. □

The following remark is obvious by the proof of Theorem 3.3.1.

Remark 3.3.2. If $(\mathcal{L} - (K_X + \Delta))\,|_{\pi^{-1}(y)}$ is ample for every $y \in Y$, then the proof of Theorem 3.3.1 shows that Theorem 3.3.1 holds over Y. This means that we can take $U = Y$ in Theorem 3.3.1.

We prove one more vanishing theorem which is historically important, although we do not need it in this book.

Theorem 3.3.3 (see [Fu4, Theorem 11.1]). *Let $\pi\colon X \to Y$ be a projective morphism of complex analytic spaces such that X is a normal complex variety, and let W be a compact subset of Y. Let Δ be an effective $\mathbb{R}$-divisor on X such that $K_X + \Delta$ is $\mathbb{R}$-Cartier. Let $\{C_i\}_{i \in I}$ be any set of log canonical centers of the pair (X, Δ). We put $V := \bigcup_{i \in I} C_i$ with the reduced structure. We assume that V is disjoint from the non-lc locus $\mathrm{Nlc}(X, \Delta)$ of (X, Δ). Let $\mathcal{M}$ be a line bundle on V such that $\mathcal{M} - (K_X + \Delta)|_V$ is π-ample over W. Then there exists some open neighborhood U of W such that $R^i\pi_*\mathcal{M} = 0$ holds on U for every $i > 0$.*

We do not pursue this direction anymore in this book. This is because it seems to be more natural to treat Theorem 3.3.3 in the framework of quasi-log complex analytic spaces. For the details of quasi-log complex analytic spaces, see [Fu16] (see also Section 6.1).

Proof of Theorem 3.3.3. As in Theorem 3.3.1, it is sufficient to prove the desired vanishing theorem for some open neighborhood of any point $w \in W$. We will use the same notation as in the proof of Theorem 3.3.1. We note that

$$A - N - (K_Z + \{\Delta_Z\} + \Delta_Z^{=1}) \sim_{\mathbb{R}} -f^*(K_X + \Delta)$$

holds. We put $f_T := f|_T : T \to V$. Then

$$f_T^* \mathcal{M} + A|_T - (K_T + (\{\Delta_Z\} + \Delta_Z^{=1} - T)|_T) \sim_{\mathbb{R}} f_T^*(\mathcal{M} - (K_X + \Delta)|_V)$$

holds. Note that $(T, (\{\Delta_Z\} + \Delta_Z^{=1} - T)|_T)$ is an analytic globally embedded simple normal crossing pair. Thus, by the vanishing theorem in Theorem 3.1.5 (2),

$$R^i\pi_*\mathcal{M} \simeq R^i\pi_*\left(\mathcal{M} \otimes (f_T)_*\mathcal{O}_T(A|_T)\right) = 0$$

for every $i > 0$. Here, we used the following isomorphism $(f|_T)_*\mathcal{O}_T(A|_T) \simeq \mathcal{O}_V$ obtained in Step 1 in the proof of Theorem 3.3.1. □

We make two remarks on Theorem 3.3.3.

Remark 3.3.4. If $(\mathcal{M} - (K_X + \Delta)|_V)|_{\pi_V^{-1}(y)}$ is ample for every $y \in \pi(V)$, where $\pi_V := \pi|_V$, then Theorem 3.3.3 holds true over Y, that is, we can take $U = Y$ in Theorem 3.3.3. We can check it by the proof of Theorems 3.3.1 and 3.3.3.

Remark 3.3.5. In [Fu4, Theorem 11.1], V is assumed to be a *minimal* log canonical center of (X, Δ) which is disjoint from $\mathrm{Nlc}(X, \Delta)$. Moreover, the proof of [Fu4, Theorem 11.1] depends on [BCHM]. For the details, see [Fu4, Remark 11.2].

We close this section with an important remark.

Remark 3.3.6. In the proof of Theorem 3.3.1, Theorem 3.1.5 for varieties is sufficient. On the other hand, the proof of Theorem 3.3.3 needs Theorem 3.1.5 (2) in full generality. Hence Theorem 3.3.3 seems to be deeper than Theorem 3.3.1. Fortunately, we do not need Theorem 3.3.3 in this book although it played a crucial role in the algebraic setting (see [Fu4]).

3.4 On log canonical centers

The main purpose of this section is to prove the following very fundamental theorem on log canonical centers, which is an easy application of Theorem 3.3.1 and its proof. It will play an important role in this book.

Theorem 3.4.1 (Basic properties of log canonical centers). *Let (X, Δ) be a log canonical pair. Then the following properties hold.*

1. *The number of log canonical centers of (X, Δ) is locally finite.*
2. *The intersection of two log canonical centers is a union of some log canonical centers.*

(3) *Let $x \in X$ be any point such that (X, Δ) is log canonical but is not kawamata log terminal at x. Then there exists a unique minimal (with respect to the inclusion) log canonical center C_x passing through x. Moreover, C_x is normal at x.*

Proof. We note that (1) is almost obvious by definition. We take an arbitrary point $x \in X$ and shrink X around x suitably. Then we may assume that there exists a projective bimeromorphic morphism $f\colon Y \to X$ from a smooth complex analytic space Y such that $K_Y + \Delta_Y := f^*(K_X + \Delta)$, $\operatorname{Supp} \Delta_Y$ is a simple normal crossing divisor on Y, and $\operatorname{Supp} \Delta_Y$ has only finitely many irreducible components (see [BieM2, Theorem 13.2]). Let $\Delta_Y^{=1} := \sum_{i \in I} \Delta_i$ be the irreducible decomposition. Then C is a log canonical center of (X, Δ) if and only if $C = f(S)$, where S is an irreducible component of $\Delta_{i_1} \cap \cdots \cap \Delta_{i_k}$ for some $\{i_1, \ldots, i_k\} \subset I$. Therefore, there exists only finitely many log canonical centers on some open neighborhood of x. Thus we obtain (1).

From now on, we will use the same notation as in the proof of Theorem 3.3.1 with $Y = X$. Let C_1 and C_2 be two log canonical centers of (X, Δ). We fix a closed point $P \in C_1 \cap C_2$. We replace X with a relatively compact Stein open neighborhood of $P \in X$ and apply the argument in the proof of Theorem 3.3.1. For the proof of (2), it is enough to find a log canonical center C such that $P \in C \subset C_1 \cap C_2$. We put $V := C_1 \cup C_2$. By Step 1 in the proof of Theorem 3.3.1, we obtain $f_*\mathcal{O}_T \simeq \mathcal{O}_V$. This means that $f\colon T \to V$ has connected fibers. We note that T is a simple normal crossing divisor on Z. Thus, there exist irreducible components T_1 and T_2 of T such that $T_1 \cap T_2 \cap f^{-1}(P) \neq \emptyset$ and that $f(T_i) \subset C_i$ for $i = 1, 2$. Therefore, we can find a log canonical center C with $P \in C \subset C_1 \cap C_2$. We finish the proof of (2). Finally, we will prove (3). The existence and the uniqueness of the minimal log canonical center follow from (2). We take the unique minimal log canonical center $C = C_x$ passing through x. We put $V := C$. We may replace X with a relatively compact Stein open neighborhood of $x \in X$. Then, by Step 1 in the proof of Theorem 3.3.1, we have $f_*\mathcal{O}_T \simeq \mathcal{O}_V$. By shrinking V around x, we can assume that every stratum of T dominates V. Let $\nu\colon V^\nu \to V$ be the normalization of V. By applying Hironaka's flattening theorem (see [Hi]) to the graph of $T \dashrightarrow V^\nu$ and then using the desingularization theorem (see [BieM2, Theorems 13.3 and 12.4]), we can obtain the following commutative diagram:

$$
\begin{array}{ccc}
T^\dagger & \xrightarrow{\ p\ } & T \\
{\scriptstyle q}\downarrow & & \downarrow{\scriptstyle f} \\
V^\nu & \xrightarrow[\ \nu\ ]{} & V,
\end{array}
$$

where $p\colon T^\dagger \to T$ is a projective bimeromorphic morphism such that $T^\dagger$ is simple normal crossing with $p_*\mathcal{O}_{T^\dagger} \simeq \mathcal{O}_T$ (see [Fu15, Lemma 2.15]). Hence

$$
\mathcal{O}_V \hookrightarrow \nu_*\mathcal{O}_{V^\nu} \hookrightarrow \nu_*q_*\mathcal{O}_{T^\dagger} \simeq f_*p_*\mathcal{O}_{T^\dagger} \simeq f_*\mathcal{O}_T \simeq \mathcal{O}_V.
$$

This implies that $\mathcal{O}_V \simeq \nu_*\mathcal{O}_{V^\nu}$ holds, that is, V is normal. Thus we obtain (3). □

By the above proof of Theorem 3.4.1, we see that Theorem 3.4.1 (2) and (3) are consequences of the strict support condition in Theorem 3.1.5 (1) through Theorem 3.3.1. We note that Theorem 3.1.5 for varieties is sufficient for Theorem 3.4.1.

Chapter 4
Fundamental theorems

In this chapter, we will prove the fundamental theorems in the theory of minimal models. In Section 4.1, we will prove the non-vanishing theorem for projective morphisms between complex analytic spaces. Our proof depends on the basepoint-free theorem for projective quasi-log schemes. Section 4.1 is completely new. Then, in Section 4.2, we will prove the basepoint-free theorem for projective morphisms between complex analytic spaces as an easy application of the non-vanishing theorem. In Section 4.3, we will establish the rationality theorem in the complex analytic setting. In Section 4.4, we will define Kleiman–Mori cones in the complex analytic setting suitably. Section 4.5 is a short section explaining Nakayama's finiteness. In Section 4.6, we will establish the cone and contraction theorem for projective morphisms between complex analytic spaces. In the final section: Section 4.7, we will prove Theorem 1.1.8 as an easy application of the vanishing theorem for projective quasi-log schemes.

4.1 Non-vanishing theorem

In this section, we will explain the non-vanishing theorem for projective morphisms between complex analytic spaces.

Theorem 4.1.1 (see [Fu4, Theorem 12.1]). *Let $\pi\colon X \to Y$ be a projective morphism of complex analytic spaces such that X is a normal complex variety, and let W be a compact subset of Y. Let Δ be an effective $\mathbb{R}$-divisor on X such that $K_X+\Delta$ is $\mathbb{R}$-Cartier. Let L be a Cartier divisor on X which is π-nef over W, that is, $L|_{\pi^{-1}(w)}$ is nef for every $w \in W$. We assume that*

(i) *$aL-(K_X+\Delta)$ is π-ample over W for some positive real number a, and*
(ii) *there exists some open neighborhood of W over which the invertible sheaf $\mathcal{O}_{\mathrm{Nlc}(X,\Delta)}(mL)$ is $\pi|_{\mathrm{Nlc}(X,\Delta)}$-generated for every $m \gg 0$.*

Then for every $m \gg 0$ there exists a relatively compact open neighborhood U_m of W over which the relative base locus $\mathrm{Bs}_\pi |mL|$ contains no log canonical centers of (X,Δ) and is disjoint from $\mathrm{Nlc}(X,\Delta)$. We note that the open subset U_m depends on m.

Remark 4.1.2. Theorem 4.1.1 is a preliminary result. Under the same assumption as in Theorem 4.1.1, we will establish the basepoint-free theorem (see Theorem 4.2.1 below), which is stronger than Theorem 4.1.1, in Section 4.2.

We first prepare the following useful lemma, which is completely new, for the proof of Theorem 4.1.1. For the details of the theory of quasi-log schemes, see [Fu6,

Chapter 6], [Fu11], and [Fu12].

Lemma 4.1.3. *Let $\pi\colon X \to Y$ be a projective morphism between complex analytic spaces such that X is a normal complex variety and let $f\colon (Z, \Delta_Z) \to X$ be a projective morphism from an analytic globally embedded simple normal crossing pair (Z, Δ_Z) such that Δ_Z is a subboundary $\mathbb{R}$-divisor on Z and it is a finite $\mathbb{R}$-linear combination of Cartier divisors, the natural map $\mathcal{O}_X \to f_*\mathcal{O}_Z(\lceil -(\Delta_Z^{<1})\rceil)$ is an isomorphism, and $K_Z + \Delta_Z \sim_{\mathbb{R}} f^*\omega$ holds for some $\mathbb{R}$-line bundle ω on X. Let y be an analytically sufficiently general point of $\pi(X)$. Then*

$$\left(X_y, \omega|_{X_y}, f_y\colon (Z_y, \Delta_{Z_y}) \to X_y\right)$$

is a projective quasi-log canonical pair, where $X_y := \pi^{-1}(y)$, $Z_y := (\pi \circ f)^{-1}(y)$, $f_y := f|_{X_y}$, and $\Delta_{Z_y} := \Delta_Z|_{Z_y}$.

This lemma is also a consequence of the strict support condition in Theorem 3.1.5 (1). We note that we need Theorem 3.1.5 (1) in full generality here.

Proof of Lemma 4.1.3. By replacing Y with $\pi(X)$, we may assume that π is surjective and Y is a complex variety. By replacing Y with a Zariski open subset of Y, we may further assume that Y is smooth. By replacing Y with a suitable Zariski open subset, we may assume that $\pi \circ f$ is flat. Then, by replacing Y with a suitable Zariski open subset again, we may assume that every stratum of $(Z, \operatorname{Supp}\Delta_Z)$ is smooth over Y. We take an arbitrary point $y \in Y$. Then (Z_y, Δ_{Z_y}) is an analytic simple normal crossing pair. From now on, we will prove that

$$\left(X_y, \omega|_{X_y}, f_y\colon (Z_y, \Delta_{Z_y}) \to X_y\right)$$

is a projective quasi-log canonical pair. Without loss of generality, we may assume that Y is a polydisc Δ^m with $y = 0 \in \Delta^m$. Let $(z_1, \dots, z_m)$ be the local coordinate system of Δ^m. Then $((\pi \circ f)^* z_i = 0)$ does not contain any strata of $(Z, \operatorname{Supp}\Delta_Z)$. Therefore, by Theorem 3.1.5 (1),

$$\pi^* z_i \times \colon R^p f_*\mathcal{O}_Z(\lceil -(\Delta_Z^{<1})\rceil) \to R^p f_*\mathcal{O}_Z(\lceil -(\Delta_Z^{<1})\rceil)$$

is injective for every i and every p since $\lceil -(\Delta_Z^{<1})\rceil - (K_Z + \{\Delta_Z\} + \Delta_Z^{=1}) \sim_{\mathbb{R}} -f^*\omega$. We put $X_1 := (\pi^* z_1 = 0)$ and $Z_1 := ((\pi \circ f)^* z_1 = 0)$. Since

$$\pi^* z_1 \times \colon R^1 f_*\mathcal{O}_Z(\lceil -(\Delta_Z^{<1})\rceil) \to R^1 f_*\mathcal{O}_Z(\lceil -(\Delta_Z^{<1})\rceil)$$

is injective, we obtain the following short exact sequence:

$$0 \longrightarrow f_*\mathcal{O}_Z(\lceil -(\Delta_Z^{<1})\rceil) \xrightarrow{\times \pi^* z_1} f_*\mathcal{O}_Z(\lceil -(\Delta_Z^{<1})\rceil) \longrightarrow f_*\mathcal{O}_{Z_1}(\lceil -(\Delta_{Z_1}^{<1})\rceil) \longrightarrow 0,$$

where $\Delta_{Z_1} = \Delta_Z|_{Z_1}$. This implies that the natural map $\mathcal{O}_{X_1} \to f_*\mathcal{O}_{Z_1}(\lceil -(\Delta_{Z_1}^{<1})\rceil)$ is an isomorphism. By repeating this argument, we finally obtain that

$$\mathcal{O}_{X_y} \simeq (f_y)_*\mathcal{O}_{Z_y}(\lceil -(\Delta_{Z_y}^{<1})\rceil)$$

holds. By [Fu7, Theorem 4.9], this means that

$$\left(X_y, \omega|_{X_y}, f_y\colon (Z_y, \Delta_{Z_y}) \to X_y\right)$$

is a projective quasi-log canonical pair. □

Let us prove Theorem 4.1.1, that is, the non-vanishing theorem.

Proof of Theorem 4.1.1. We divide the proof into several small steps.

Step 1. By shrinking Y suitably, we may assume that there exists a positive integer m_1 such that $\mathcal{O}_{\mathrm{Nlc}(X,\Delta)}(mL)$ is $\pi|_{\mathrm{Nlc}(X,\Delta)}$-generated for every $m \geq m_1$ by (ii). We may further assume that $aL - (K_X + \Delta)$ is π-ample over Y.

Step 2. In this step, we will prove the following claim.

Claim. *There exists a positive integer m_2 such that $\pi_*\mathcal{O}_V(mL) \neq 0$ holds for every $m \geq m_2$, where V is any minimal log canonical center of (X, Δ) such that $\pi(V) \cap W \neq \emptyset$ and that $V \cap \mathrm{Nlc}(X, \Delta) = \emptyset$ over some open neighborhood of W.*

Proof of Claim. We note that the number of the minimal log canonical centers V of (X, Δ) with $\pi(V) \cap W \neq \emptyset$ is finite. We take a minimal log canonical center V such that $\pi(V) \cap W \neq \emptyset$ and that $V \cap \mathrm{Nlc}(X, \Delta) = \emptyset$ over some open neighborhood of W. Let y be an arbitrary point of $\pi(V) \cap W$. It is sufficient to prove $\pi_*\mathcal{O}_V(mL) \neq 0$ on a small open neighborhood of y. Therefore, we can replace Y with a small relatively compact Stein open neighborhood of y. Thus, by Step 1 in the proof of Theorem 3.3.1, we can construct a projective surjective morphism $f\colon (T, \Delta_T) \to V$ from an analytic globally embedded simple normal crossing pair (T, Δ_T) such that Δ_T is a subboundary $\mathbb{R}$-divisor on T, the natural map $\mathcal{O}_V \to f_*\mathcal{O}_T(\lceil -(\Delta_T^{<1}) \rceil)$ is an isomorphism, and $K_T + \Delta_T \sim_{\mathbb{R}} f^*(K_X + \Delta)|_V$ holds. Thus, by Lemma 4.1.3, an analytically sufficiently general fiber F of $\pi\colon V \to \pi(V)$ is a projective quasi-log canonical pair. By Lemma 2.2.5, we may assume that $L|_F$ is nef. Therefore, by the basepoint-free theorem for quasi-log canonical pairs (see [Fu6, Theorem 6.5.1]), there exists a positive integer m_2 such that $|mL|_F|$ is basepoint-free for every $m \geq m_2$. This implies that $\pi_*\mathcal{O}_V(mL) \neq 0$ for every $m \geq m_2$. This is what we wanted. □

Step 3. We put $m_0 := \max\{a, m_1, m_2\}$. Let m be any positive integer with $m \geq m_0$. Since $aL - (K_X + \Delta)$ is π-ample over W and L is π-nef over W, $mL - (K_X + \Delta)$ is π-ample over W. By Theorem 3.3.1, we can find an open neighborhood U_m of W such that the vanishing theorem holds for mL over U_m. Without loss of generality, we may assume that every minimal log canonical center V of (X, Δ) with $\pi(V) \cap U_m \neq \emptyset$ always satisfies $\pi(V) \cap W \neq \emptyset$ by shrinking U_m suitably.

Step 4. In this final step, we will prove that over U_m the relative base locus $\mathrm{Bs}_\pi |mL|$ contains no log canonical centers of (X, Δ) and is disjoint from $\mathrm{Nlc}(X, \Delta)$ for every $m \geq m_0$.

By the vanishing theorem (see Theorem 3.3.1), we have

$$R^1\pi_*\left(\mathcal{J}_{\mathrm{NLC}}(X,\Delta)\otimes\mathcal{O}_X(mL)\right)=0$$

on U_m. Thus the restriction map

$$\pi_*\mathcal{O}_X(mL)\to\pi_*\mathcal{O}_{\mathrm{Nlc}(X,\Delta)}(mL)$$

is surjective on U_m. This implies that the relative base locus $\mathrm{Bs}_\pi|mL|$ is disjoint from $\mathrm{Nlc}(X,\Delta)$ over U_m. Let V be a minimal log canonical center of (X,Δ) with $\pi(V)\cap U_m\neq\emptyset$. If $V\cap\mathrm{Nlc}(X,\Delta)\neq\emptyset$ over U_m, then $V\not\subset\mathrm{Bs}_\pi|mL|$ since $\mathrm{Nlc}(X,\Delta)\cap\mathrm{Bs}_\pi|mL|=\emptyset$ over U_m. Hence we may assume that $V\cap\mathrm{Nlc}(X,\Delta)=\emptyset$ over U_m. In this case, $\pi_*\mathcal{O}_V(mL)\neq 0$ by Claim in Step 2. On the other hand, by the vanishing theorem (see Theorem 3.3.1), the restriction map

$$\pi_*\mathcal{O}_X(mL)\to\pi_*\mathcal{O}_V(mL)$$

is surjective on U_m. This implies that $V\not\subset\mathrm{Bs}_\pi|mL|$. Hence $\mathrm{Bs}_\pi|mL|$ contains no log canonical centers of (X,Δ) over U_m.

We finish the proof. □

We make an important remark on Theorem 4.1.1.

Remark 4.1.4. In Step 3 in the proof of Theorem 4.1.1, the condition that $mL-(K_X+\Delta)$ is π-ample over $w\in W$ only implies that $mL-(K_X+\Delta)$ is π-ample over some open neighborhood $\mathcal{U}^m_w$ of w in Y. We note that $\mathcal{U}^m_w$ depends on m. Therefore, U_m in Theorem 4.1.1 also depends on m.

For kawamata log terminal pairs, the non-vanishing theorem is formulated as follows. We note that Theorem 4.1.1 is vacuous when (X,Δ) is kawamata log terminal.

Theorem 4.1.5 (Non-vanishing theorem for kawamata log terminal pairs). *Let $\pi\colon X\to Y$ be a projective morphism of complex analytic spaces such that X is a normal complex variety and let W be a compact subset of Y. Let Δ be an effective $\mathbb{R}$-divisor on X such that (X,Δ) is kawamata log terminal. Let L be a Cartier divisor on X which is π-nef over W. We assume that $aL-(K_X+\Delta)$ is π-ample over W for some positive real number a. Then $\pi_*\mathcal{O}_X(mL)\neq 0$ holds for every $m\gg 0$*

Proof. By shrinking Y around W suitably, we may assume that $aL-(K_X+\Delta)$ is π-ample over Y. Let F be an analytically sufficiently general fiber of $\pi\colon X\to\pi(X)$. Then $(F,\Delta|_F)$ is kawamata log terminal. By Lemma 2.2.5, we may assume that $L|_F$ is nef. Hence $|mL|_F|$ is basepoint-free for every $m\gg 0$ by the usual Kawamata–Shokurov basepoint-free theorem for projective kawamata log terminal pairs. Thus, we obtain that $\pi_*\mathcal{O}_X(mL)\neq 0$ for every $m\gg 0$. This is what we wanted. □

We will use Theorems 4.1.1 and 4.1.5 in the proof of the basepoint-freeness in Section 4.2.

4.2 Basepoint-free theorem

In this section, we will explain the basepoint-free theorem in the complex analytic setting, which is new even when $\pi\colon (X,\Delta) \to Y$ is algebraic (see Theorem 1.3.6).

Theorem 4.2.1 (see [Fu4, Theorem 13.1]). *Let $\pi\colon X \to Y$ be a projective morphism of complex analytic spaces such that X is a normal complex variety and let W be a compact subset of Y. Let Δ be an effective $\mathbb{R}$-divisor on X such that $K_X+\Delta$ is $\mathbb{R}$-Cartier. Let L be a Cartier divisor on X which is π-nef over W. We assume that*

(i) *$aL-(K_X+\Delta)$ is π-ample over W for some positive real number a, and*

(ii) *there exists some open neighborhood of W over which the invertible sheaf $\mathcal{O}_{\mathrm{Nlc}(X,\Delta)}(mL)$ is $\pi|_{\mathrm{Nlc}(X,\Delta)}$-generated for every $m \gg 0$.*

Then there exists a relatively compact open neighborhood U of W such that $\mathcal{O}_X(mL)$ is π-generated over U for every $m \gg 0$.

Theorem 4.2.1 is a consequence of the non-vanishing theorem (see Theorems 4.1.1 and 4.1.5).

Proof of Theorem 4.2.1. We take an arbitrary point $y \in W$. It is sufficient to prove that $\mathcal{O}_X(mL)$ is π-generated for every $m \gg 0$ over some relatively compact Stein open neighborhood of y. Hence, we will sometimes freely replace Y with a suitable relatively compact Stein open neighborhood of y without mentioning it explicitly throughout this proof. So, from now on, we assume that Y is Stein and that π is surjective.

Step 1. Let p be a prime number. In this step, we will prove that there exists a positive integer k such that $\mathrm{Bs}_\pi |p^kL| = \emptyset$ holds over some open neighborhood of y, where $\mathrm{Bs}_\pi |p^kL| := \mathrm{Supp}\left(\mathrm{Coker}(\pi^*\pi_*\mathcal{O}_X(p^kL) \to \mathcal{O}_X(p^kL))\right)$.

By putting $W := \{y\}$ and using the non-vanishing theorem (see Theorems 4.1.1 and 4.1.5), we obtain $|p^{k_1}L| \neq \emptyset$ for some positive integer k_1. If $\mathrm{Bs}_\pi |p^{k_1}L| = \emptyset$, then there is nothing to prove. Hence we may assume that $\mathrm{Bs}_\pi |p^{k_1}L| \neq \emptyset$. By Theorem 4.1.1, we may assume that $\mathrm{Bs}_\pi |p^{k_1}L|$ contains no log canonical centers of (X,Δ) and is disjoint from $\mathrm{Nlc}(X,\Delta)$ after shrinking Y suitably. By shrinking Y around $W=\{y\}$, we may assume that $\pi(V)\cap W \neq \emptyset$, where V is any irreducible component of $\mathrm{Bs}_\pi |p^{k_1}L|$. Without loss of generality, we may further assume that $aL-(K_X+\Delta)$ is π-ample over Y. We take general members $D_1, \ldots, D_{n+1}$ of $|p^{k_1}L|$ with $n=\dim X$. We put $D := \sum_{i=1}^{n+1} D_i$. We may assume that $(X,\Delta+D)$ is log canonical outside $\mathrm{Bs}_\pi |p^{k_1}L| \cup \mathrm{Nlc}(X,\Delta)$. Let $x \in X$ be any point of $\mathrm{Bs}_\pi |p^{k_1}L|$. Then, by Lemma 2.1.3, $(X,\Delta+D)$ is not log canonical at x. We put

$$c := \sup\left\{t \in \mathbb{R} \mid (X,\Delta+tD) \text{ is log canonical at } \pi^{-1}(y)\cap(X\setminus \mathrm{Nlc}(X,\Delta))\right\}.$$

Then we can check that $0<c<1$. By shrinking Y around $W=\{y\}$ suitably again, we may assume that $(X,\Delta+cD)$ is log canonical outside $\mathrm{Nlc}(X,\Delta)$. By Lemma

2.3.3 and Remark 2.3.4, we see that $\mathcal{J}_{\mathrm{NLC}}(X, \Delta + cD) = \mathcal{J}_{\mathrm{NLC}}(X, \Delta)$ holds. By construction,

$$\left(c(n+1)p^{k_1} + a\right) L - (K_X + \Delta + cD) \sim_{\mathbb{R}} aL - (K_X + \Delta)$$

is π-ample over Y. By construction again, there exists a log canonical center V of $(X, \Delta + cD)$ which is contained in $\mathrm{Bs}_\pi |p^{k_1}L|$ such that $\pi(V) \cap W \neq \emptyset$. By the non-vanishing theorem (see Theorem 4.1.1), we can find $k_2 > k_1$ such that $\mathrm{Bs}_\pi |p^{k_2}L| \subsetneq \mathrm{Bs}_\pi |p^{k_1}L|$. By repeating this process finitely many times, we can find k such that $\mathrm{Bs}_\pi |p^k L| = \emptyset$ over some open neighborhood of y by the noetherian induction (see [Fi, 0.40. Corollary]).

Step 2. We take another prime number p'. Then there exists a positive integer k' such that $\mathrm{Bs}_\pi |p'^{k'} L| = \emptyset$ over some open neighborhood of y by Step 1. This implies that there exist a positive integer m_0 and some open neighborhood U_y of y such that for every $m \geq m_0$ the relative base locus $\mathrm{Bs}_\pi |mL|$ is empty over U_y.

Since W is compact, we obtain a desired open neighborhood U of W. We finish the proof. □

Remark 4.2.2. Although the non-vanishing theorem (see Theorems 4.1.1 and 4.1.5) and the basepoint-free theorem (see Theorem 4.2.1) were formulated for Cartier divisors L, it is obvious that they hold true even for line bundles $\mathcal{L}$. We will sometimes use the basepoint-free theorem for line bundles in subsequent sections. As we explained in Section 1.3, there are many complex manifolds which have no non-trivial Cartier divisors (see Example 1.3.3). Hence it may be better to formulate Theorems 4.1.1, 4.1.5, 4.2.1, and so on, for line bundles than for Cartier divisors.

In the proof of Theorem 4.1.1, we used the basepoint-free theorem for projective quasi-log schemes (see [Fu6, Theorem 6.5.1]). Hence the proof of Theorem 4.2.1 in this book does not give a new proof of the basepoint-free theorem of normal pairs in the algebraic setting (see [Fu4, Theorem 13.1]).

4.3 Rationality theorem

In this section, we will explain the rationality theorem in the complex analytic setting. Although the proof of [Fu4, Theorem 15.1], which is the rationality theorem in the algebraic setting, works with some suitable modifications, we will explain the details for the reader's convenience. This is because the proof of the rationality theorem is complicated.

Theorem 4.3.1 (Rationality theorem, see [Fu4, Theorem 15.1]). *Let $\pi\colon X \to Y$ be a projective morphism of complex analytic spaces such that X is a normal complex variety, and let W be a compact subset of Y. Let Δ be an effective $\mathbb{Q}$-divisor on X such that $K_X + \Delta$ is $\mathbb{Q}$-Cartier. Let H be a π-ample Cartier divisor on X. Assume that $K_X + \Delta$ is not π-nef over W and that r is a positive real number such that*

(i) *$H + r(K_X + \Delta)$ is π-nef over W but is not π-ample over W, and*

(ii) $(H + r(K_X + \Delta))|_{\mathrm{Nlc}(X,\Delta)}$ *is* $\pi|_{\mathrm{Nlc}(X,\Delta)}$*-ample over* W.

Then r *is a rational number, and in reduced form, it has denominator at most* $a(d+1)$, *where* $d := \max_{w\in W}\dim \pi^{-1}(w)$ *and* a *is a positive integer such that* $a(K_X + \Delta)$ *is a Cartier divisor in a neighborhood of* $\pi^{-1}(W)$.

In the proof of Theorem 4.3.1, we will use the following elementary lemmas.

Lemma 4.3.2 ([KM, Lemma 3.19]). *Let* $P(x, y)$ *be a non-trivial polynomial of degree* $\leq d$ *and assume that* P *vanishes for all sufficiently large integral solutions of* $0 < ay - rx < \varepsilon$ *for some fixed positive integer* a *and positive* ε *for some* $r \in \mathbb{R}$. *Then* r *is rational, and in reduced form,* r *has denominator* $\leq a(d+1)/\varepsilon$.

We give a proof of Lemma 4.3.2 for the sake of completeness.

Proof of Lemma 4.3.2. We first assume that r is irrational. Then an infinite number of integral points in the (x, y)-plane on each side of the line $ay - rx = 0$ are closer than $\varepsilon/(d+2)$ to that line. Hence there is a large integral solution (x', y') with $0 < ay' - rx' < \varepsilon/(d+2)$. In this case, we see that

$$(2x', 2y'), \cdots, ((d+1)x', (d+1)y')$$

are solutions by hypothesis. Thus $(y'x - x'y)$ divides P since P and $(y'x - x'y)$ have $(d+1)$ common zeros. We choose a smaller ε and repeat the argument. We do this $d+1$ times to get a contradiction. Therefore, r is a rational number.

Now we assume that $r = u/v$ in lowest terms. For given j, let (x', y') be a solution of $ay - rx = aj/v$. Note that an integral solution exists for every j. Then we have

$$a(y' + kv) - r(x' + akv) = aj/v$$

for all k. Hence, as above, if $aj/v < \varepsilon$, $(ay - rx) - (aj/v)$ must divide P. Therefore, we can have at most d such values of j. Thus $a(d+1)/v \geq \varepsilon$. □

Lemma 4.3.3. *Let* F *be a projective variety and let* D_1 *and* D_2 *be Cartier divisors on* X. *Let us consider the Hilbert polynomial*

$$P(u_1, u_2) := \chi(F, \mathcal{O}_F(u_1D_1 + u_2D_2)).$$

If D_1 *is ample, then* $P(u_1, u_2)$ *is a non-trivial polynomial of total degree* $\leq \dim F$. *It is because* $P(u_1, 0) = \dim_{\mathbb{C}} H^0(F, \mathcal{O}_F(u_1D_1)) \not\equiv 0$ *if* u_1 *is sufficiently large.*

Let us start the proof of Theorem 4.3.1.

Proof of Theorem 4.3.1. Throughout this proof, we can freely shrink Y around W suitably. Hence we will sometimes replace Y with a small open neighborhood of W without mentioning it explicitly.

Let m be a positive integer such that $H' := mH$ is π-very ample after shrinking Y around W suitably. If $H' + r'(K_X + \Delta)$ is π-nef over W but is not π-ample over W, and $(H' + r'(K_X + \Delta))|_{\mathrm{Nlc}(X,\Delta)}$ is $\pi|_{\mathrm{Nlc}(X,\Delta)}$-ample over W, then

$$H + r(K_X + \Delta) = \frac{1}{m}(H' + r'(K_X + \Delta))$$

holds. This implies that $r = (1/m)r'$ holds. Therefore, it is obvious that r is rational if and only if r' is rational. We further assume that r' has denominator v. Then r has denominator dividing mv. Since m can be an arbitrary sufficiently large positive integer, this means that r has denominator dividing v. Hence, by replacing H with mH, we may assume that H is π-very ample.

For each $(p, q) \in \mathbb{Z}^2$, we put $M(p, q) := pH + qa(K_X + \Delta)$ and define

$$L(p, q) := \operatorname{Supp}(\operatorname{Coker}(\pi^*\pi_*\mathcal{O}_X(M(p, q)) \to \mathcal{O}_X(M(p, q)))).$$

By definition, $L(p, q) = X$ holds if and only if $\pi_*\mathcal{O}_X(M(p, q)) = 0$.

Claim 1. *Let ε be a positive number. For (p, q) sufficiently large and $0 < aq - rp < \varepsilon$, $L(p, q)$ is the same subset of X after shrinking Y around W suitably. We call this subset L_0. Let $I \subset \mathbb{Z}^2$ be the set of (p, q) for which $0 < aq - rp < 1$ and $L(p, q) = L_0$. Then we note that I contains all sufficiently large (p, q) with $0 < aq - rp < 1$.*

Proof of Claim 1. We fix $(p_0, q_0) \in \mathbb{Z}^2$ such that $p_0 > 0$ and $0 < aq_0 - rp_0 < 1$. Since H is π-very ample, there exists a positive integer m_0 such that $\mathcal{O}_X(mH + ja(K_X + \Delta))$ is π-generated for every $m > m_0$ and every $0 \leq j \leq q_0 - 1$ after shrinking Y around W suitably. Let M be the round-up of

$$\left(m_0 + \frac{1}{r}\right) \Big/ \left(\frac{a}{r} - \frac{p_0}{q_0}\right).$$

If $(p', q') \in \mathbb{Z}^2$ such that $0 < aq' - rp' < 1$ and $q' \geq M + q_0 - 1$, then we can write

$$p'H + q'a(K_X + \Delta) = k(p_0H + q_0a(K_X + \Delta)) + (lH + ja(K_X + \Delta))$$

for some $k \geq 0$, $0 \leq j \leq q_0 - 1$ with $l > m_0$. It is because we can uniquely write $q' = kq_0 + j$ with $0 \leq j \leq q_0 - 1$. Thus, we have $kq_0 \geq M$. So, we obtain

$$l = p' - kp_0 > \frac{a}{r}q' - \frac{1}{r} - (kq_0)\frac{p_0}{q_0} \geq \left(\frac{a}{r} - \frac{p_0}{q_0}\right) M - \frac{1}{r} \geq m_0.$$

Therefore, $L(p', q') \subset L(p_0, q_0)$. We note that we can use the noetherian induction over a relatively compact open neighborhood of W (see [Fi, 0.40. Corollary]). Therefore, after shrinking Y around W suitably again, we obtain the desired closed subset $L_0 \subset X$. We can check that the subset $I \subset \mathbb{Z}^2$ contains all sufficiently large (p, q) with $0 < aq - rp < 1$ without any difficulties. □

Claim 2. *We have $L_0 \cap \operatorname{Nlc}(X, \Delta) = \emptyset$.*

Proof of Claim 2. We take $(\alpha, \beta) \in \mathbb{Q}^2$ such that $\alpha > 0$, $\beta > 0$, and $\beta a/\alpha > r$ is sufficiently close to r. Then $(\alpha H + \beta a(K_X + \Delta))|_{\operatorname{Nlc}(X,\Delta)}$ is $\pi|_{\operatorname{Nlc}(X,\Delta)}$-ample over W because $(H + r(K_X + \Delta))|_{\operatorname{Nlc}(X,\Delta)}$ is $\pi|_{\operatorname{Nlc}(X,\Delta)}$-ample over W. We take any point $w \in W$. Then it is sufficient to prove that $L_0 \cap \operatorname{Nlc}(X, \Delta) = \emptyset$ holds over some open neighborhood of w. From now on, we will freely shrink Y around w

without mentioning it explicitly. We take a sufficiently large and divisible positive integer m' such that

$$m'(\alpha H + \beta a(K_X + \Delta))|_{\mathrm{Nlc}(X,\Delta)}$$

is $\pi|_{\mathrm{Nlc}(X,\Delta)}$-very ample. We put $(p_0, q_0) := (m'\alpha, m'\beta)$ and apply the argument in the proof of Claim 1. Thus, if $0 < aq - rp < 1$ and $(p, q) \in \mathbb{Z}^2$ is sufficiently large, then we can write

$$M(p,q) = mM(\alpha,\beta) + (M(p,q) - mM(\alpha,\beta))$$

such that $M(p,q) - mM(\alpha,\beta)$ is π-very ample and that

$$m(\alpha H + \beta a(K_X + \Delta))|_{\mathrm{Nlc}(X,\Delta)}$$

is also $\pi|_{\mathrm{Nlc}(X,\Delta)}$-very ample. Hence, $\mathcal{O}_{\mathrm{Nlc}(X,\Delta)}(M(p,q))$ is π-very ample. We note that

$$M(p,q) - (K_X + \Delta) = pH + (qa - 1)(K_X + \Delta)$$

is π-ample over some open neighborhood of w because (p,q) is sufficiently large and $aq - rp < 1$. Thus, by the vanishing theorem: Theorem 3.3.1, the restriction map

$$\pi_*\mathcal{O}_X(M(p,q)) \to \pi_*\mathcal{O}_{\mathrm{Nlc}(X,\Delta)}(M(p,q))$$

is surjective. Therefore, $L(p,q) \cap \mathrm{Nlc}(X,\Delta) = \emptyset$ holds over some open neighborhood of w. By Claim 1, we have $L_0 \cap \mathrm{Nlc}(X,\Delta) = \emptyset$ over some open neighborhood of w. Since w is an arbitrary point of W, $L_0 \cap \mathrm{Nlc}(X,\Delta) = \emptyset$ holds over some open neighborhood of W. This is what we wanted. □

Claim 3. *We assume that r is not rational or that r is rational and has denominator $> a(d+1)$ in reduced form. Then, for (p,q) sufficiently large and $0 < aq - rp < 1$, $\mathcal{O}_X(M(p,q))$ is π-generated at general points of every log canonical center of (X,Δ).*

We will explain the proof of Claim 3 in detail because we have to change the proof of Claim 3 in the proof of [Fu4, Theorem 15.1] slightly.

Proof of Claim 3. After shrinking Y around W suitably, it is sufficient to consider minimal log canonical centers C of (X,Δ) such that $\pi(C) \cap W \neq \emptyset$. By Claim 2, we may assume that $C \cap \mathrm{Nlc}(X,\Delta) = \emptyset$ holds. We take a point $w \in \pi(C) \cap W$. It is sufficient to consider everything over some small open neighborhood of w in Y. We take an analytically sufficiently general fiber F of $C \to \pi(C)$. Then we may assume that $(H + r(K_X + \Delta))|_F$ and $(H + r(K_X + \Delta))|_{\pi^{-1}(\pi(F))}$ are both nef by Lemma 2.2.5 (see also Remark 2.2.7). We note that

$$\begin{aligned} M(p,q) - (K_X + \Delta) &= pH + (qa - 1)(K_X + \Delta) \\ &= \left(p - \frac{qa-1}{r}\right)H + \frac{(qa-1)}{r}(H + r(K_X + \Delta)) \end{aligned}$$

holds. Therefore, if $aq - rp < 1$ and (p,q) is sufficiently large, then we see that

$$(M(p,q) - (K_X + \Delta))|_{\pi^{-1}(\pi(F))}$$

is ample. We note that

$$P_F(p,q) := \chi(F, \mathcal{O}_F(M(p,q)))$$

is a non-zero polynomial of degree at most $\dim F \leq d$ by Lemma 4.3.3. We also note that F is an analytically sufficiently general fiber of $C \to \pi(C)$. By Lemma 4.3.2, there exists (p,q) such that $P_F(p,q) \neq 0$, (p,q) sufficiently large, and $0 < aq - rp < 1$. By the π-ampleness of $M(p,q) - (K_X + \Delta)$ over some open neighborhood of $\pi(F)$,

$$P_F(p,q) = \chi(F, \mathcal{O}_F(M(p,q))) = \dim_{\mathbb{C}} H^0(F, \mathcal{O}_F(M(p,q)))$$

and

$$\pi_*\mathcal{O}_X(M(p,q)) \to \pi_*\mathcal{O}_C(M(p,q))$$

is surjective over some open neighborhood of $\pi(F)$ by Theorem 3.3.1 (see also 2.1.14). We note that $\pi_*\mathcal{O}_C(M(p,q)) \neq 0$ by $P_F(p,q) \neq 0$ and that $C \cap \mathrm{Nlc}(X,\Delta) = \emptyset$ by assumption. Therefore, $\mathcal{O}_X(M(p,q))$ is π-generated at general points of C. By combining this fact with Claim 1, $\mathcal{O}_X(M(p,q))$ is π-generated at general points of every log canonical center of (X,Δ) if (p,q) is sufficiently large with $0 < aq - rp < 1$. Hence we obtain Claim 3. □

Note that $\mathcal{O}_X(M(p,q))$ is not π-generated for $(p,q) \in I$ because $M(p,q)$ is not π-nef over W. Therefore, $L_0 \neq \emptyset$ with $\pi(L_0) \cap W \neq \emptyset$. We take a point $w \in \pi(L_0) \cap W$ and replace Y with a relatively compact Stein open neighborhood of w. From now on, we will freely shrink Y around w suitably. Let $D_1, \ldots, D_{n+1}$ be general members of

$$\pi_*\mathcal{O}_X(M(p_0,q_0)) = H^0(X, \mathcal{O}_X(M(p_0,q_0)))$$

with $(p_0,q_0) \in I$. We put $D := \sum_{i=1}^{n+1} D_i$. Let $x \in X$ be any point of L_0. Then, by Lemma 2.1.3, $K_X + \Delta + D$ is not log canonical at x. On the other hand, we may assume that $K_X + \Delta + D$ is log canonical outside $L_0 \cup \mathrm{Nlc}(X,\Delta)$ since D_i is a general member of $|M(p_0,q_0)|$ for every i. We put

$$c := \sup\left\{t \in \mathbb{R} \mid (X, \Delta + tD) \text{ is log canonical at } \pi^{-1}(w) \cap (X \setminus \mathrm{Nlc}(X,\Delta))\right\}.$$

Then we can check that $0 < c < 1$ by Claim 3. We note that $w \in \pi(L_0) \cap W$. Thus, the pair $(X, \Delta + cD)$ has some log canonical centers contained in L_0 and intersecting $\pi^{-1}(w)$. By shrinking Y around w, we may assume that $(X, \Delta + cD)$ is log canonical outside $\mathrm{Nlc}(X,\Delta)$. Let C be a log canonical center contained in L_0 and intersecting $\pi^{-1}(w)$. We note that

$$\mathcal{J}_{\mathrm{NLC}}(X, \Delta + cD) = \mathcal{J}_{\mathrm{NLC}}(X,\Delta)$$

by Lemma 2.3.3 and Remark 2.3.4 and that

$$C \cap \mathrm{Nlc}(X, \Delta + cD) = C \cap \mathrm{Nlc}(X,\Delta) = \emptyset.$$

We consider

$$K_X + \Delta + cD = c(n+1)p_0 H + (1 + c(n+1)q_0 a)(K_X + \Delta).$$

Thus we have

$$\begin{aligned}&pH+qa(K_X+\Delta)-(K_X+\Delta+cD)\\&=(p-c(n+1)p_0)H+(qa-(1+c(n+1)q_0a))(K_X+\Delta).\end{aligned}$$

If p and q are large enough and $0<aq-rp\le aq_0-rp_0$, then

$$pH+qa(K_X+\Delta)-(K_X+\Delta+cD)$$

is π-ample over w. It is because

$$\begin{aligned}&(p-c(n+1)p_0)H+(qa-(1+c(n+1)q_0a))(K_X+\Delta)\\&=(p-(1+c(n+1))p_0)H+(qa-(1+c(n+1))q_0a)(K_X+\Delta)\\&+p_0H+(q_0a-1)(K_X+\Delta).\end{aligned}$$

By shrinking Y around w suitably, we may further assume that it is π-ample over Y. We consider an analytically sufficiently general fiber F of $C\to\pi(C)$ as in the proof of Claim 3. We note that $(H+r(K_X+\Delta))|_{\pi^{-1}(\pi(F))}$ is nef by the choice of F.

Suppose that r is not rational. There exists an arbitrarily large $(p,q)\in\mathbb{Z}^2$ such that $0<aq-rp<\varepsilon=aq_0-rp_0$ and $\chi(F,\mathcal{O}_F(M(p,q)))\neq 0$ by Lemma 4.3.2 because $P_F(p,q)=\chi(F,\mathcal{O}_F(M(p,q)))$ is a non-trivial polynomial of degree at most $\dim F\le d$ by Lemma 4.3.3. Since

$$(M(p,q)-(K_X+\Delta+cD))|_{\pi^{-1}(\pi(F))}$$

is ample by $0<aq-rp<aq_0-rp_0$, we have

$$\dim_{\mathbb{C}}H^0(F,\mathcal{O}_F(M(p,q)))=\chi(F,\mathcal{O}_F(M(p,q)))\neq 0$$

by the vanishing theorem: Theorem 3.3.1 (see also 2.1.14). By the vanishing theorem: Theorem 3.3.1,

$$\pi_*\mathcal{O}_X(M(p,q))\to\pi_*\mathcal{O}_C(M(p,q))$$

is surjective over a neighborhood of $\pi(F)$ because

$$(M(p,q)-(K_X+\Delta+cD))|_{\pi^{-1}(\pi(F))}$$

is ample. We note that $C\cap\mathrm{Nlc}(X,\Delta+cD)=\emptyset$. Thus C is not contained in $L(p,q)$. Therefore, $L(p,q)$ is a proper subset of $L(p_0,q_0)=L_0$, which gives the desired contradiction. Hence we know that r is rational.

We next suppose that the assertion of the theorem concerning the denominator of r is false. We choose $(p_0,q_0)\in I$ such that aq_0-rp_0 is the maximum, say it is equal to e/v. If $0<aq-rp\le e/v$ and (p,q) is sufficiently large, then

$$\chi(F,\mathcal{O}_F(M(p,q)))=\dim_{\mathbb{C}}H^0(F,\mathcal{O}_F(M(p,q)))$$

since $(M(p,q)-(K_X+\Delta+cD))|_{\pi^{-1}(\pi(F))}$ is ample. There exists sufficiently large $(p,q)\in\mathbb{Z}^2$ in the strip $0<aq-rp<1$ with $\varepsilon=1$ for which

$$\dim_{\mathbb{C}} H^0(F,\mathcal{O}_F(M(p,q)))=\chi(F,\mathcal{O}_F(M(p,q)))\neq 0$$

by Lemma 4.3.2 since $P_F(p,q)=\chi(F,\mathcal{O}_F(M(p,q)))$ is a non-trivial polynomial of degree at most $\dim F\leq d$ by Lemma 4.3.3. Note that $aq-rp\leq e/v=aq_0-rp_0$ holds automatically for $(p,q)\in I$. Since

$$\pi_*\mathcal{O}_X(M(p,q))\to\pi_*\mathcal{O}_C(M(p,q))$$

is surjective over some open neighborhood of $\pi(F)$ by the ampleness of

$$(M(p,q)-(K_X+\Delta+cD))|_{\pi^{-1}(\pi(F))},$$

we obtain the desired contradiction by the same reason as above.

Thus, we finish the proof of the rationality theorem. □

We close this section with an easy remark.

Remark 4.3.4. The proof of Theorem 4.3.1 shows that Theorem 4.3.1 holds true under the assumption that H is a π-ample line bundle.

4.4 Kleiman–Mori cones

In this section, we will define Kleiman–Mori cones for projective morphisms between complex analytic spaces under some suitable assumption.

4.4.1. Throughout this section, let $\pi\colon X\to Y$ be a projective morphism of complex analytic spaces and let W be a compact subset of Y. Let $Z_1(X/Y;W)$ be the free abelian group generated by the projective integral curves C on X such that $\pi(C)$ is a point of W. Let U be any open neighborhood of W. Then we can consider the following intersection pairing

$$\cdot\colon \mathrm{Pic}\big(\pi^{-1}(U)\big)\times Z_1(X/Y;W)\to\mathbb{Z}$$

given by $\mathcal{L}\cdot C\in\mathbb{Z}$ for $\mathcal{L}\in\mathrm{Pic}\big(\pi^{-1}(U)\big)$ and $C\in Z_1(X/Y;W)$. We say that $\mathcal{L}$ is *π-numerically trivial over W* when $\mathcal{L}\cdot C=0$ for every $C\in Z_1(X/Y;W)$. We take $\mathcal{L}_1,\mathcal{L}_2\in\mathrm{Pic}\big(\pi^{-1}(U)\big)$. If $\mathcal{L}_1\otimes\mathcal{L}_2^{-1}$ is π-numerically trivial over W, then we write $\mathcal{L}_1\equiv_W\mathcal{L}_2$ and say that $\mathcal{L}_1$ is numerically equivalent to $\mathcal{L}_2$ over W. We put

$$\widetilde{A}(U,W):=\mathrm{Pic}\big(\pi^{-1}(U)\big)/{\equiv_W}$$

and define

$$A^1(X/Y;W):=\varinjlim_{W\subset U}\widetilde{A}(U,W),$$

where U runs through all the open neighborhoods of W.

4.4.2. We assume that $A^1(X/Y;W)$ is a finitely generated abelian group. Then we can define the *relative Picard number* $\rho(X/Y;W)$ to be the rank of $A^1(X/Y;W)$. We put

$$N^1(X/Y;W) := A^1(X/Y;W) \otimes_{\mathbb{Z}} \mathbb{R}.$$

Let $A_1(X/Y;W)$ be the image of

$$Z_1(X/Y;W) \to \operatorname{Hom}_{\mathbb{Z}}\left(A^1(X/Y;W), \mathbb{Z}\right)$$

given by the above intersection pairing. Then we set

$$N_1(X/Y;W) := A_1(X/Y;W) \otimes_{\mathbb{Z}} \mathbb{R}.$$

As usual, we can define the *Kleiman–Mori cone*

$$\overline{\mathrm{NE}}(X/Y;W)$$

of $\pi\colon X \to Y$ over W, that is, $\overline{\mathrm{NE}}(X/Y;W)$ is the closure of the convex cone in $N_1(X/Y;W)$ spanned by the projective integral curves C on X such that $\pi(C)$ is a point of W. An element $\zeta \in N^1(X/Y;W)$ is called *π-nef over W* or *nef over W* if $\zeta \geq 0$ on $\overline{\mathrm{NE}}(X/Y;W)$, equivalently, $\zeta|_{\pi^{-1}(w)}$ is nef in the usual sense for every $w \in W$.

Remark 4.4.3. We assume that $\pi\colon X \to Y$ decomposes as

$$\pi\colon X \xrightarrow{\varphi} Z \xrightarrow{\pi_Z} Y,$$

where $\pi_Z\colon Z \to Y$ is a projective morphism of complex analytic spaces. Then φ is always projective and $\pi_Z^{-1}(W)$ is a compact subset of Z. Therefore, we can define $A^1\left(X/Z;\pi_Z^{-1}(W)\right)$ and $N^1\left(X/Z;\pi_Z^{-1}(W)\right)$ as above. By definition, $N^1\left(X/Z;\pi_Z^{-1}(W)\right)$ is a quotient vector space of $N^1\left(X/Y;W\right)$. Hence, if the dimension of $N^1(X/Y;W)$ is finite, then we see that $\dim_{\mathbb{R}} N^1\left(X/Z;\pi_Z^{-1}(W)\right) < \infty$ holds.

Lemma 4.4.4 ([Na1, Proposition 4.7 (2)]). *$\overline{\mathrm{NE}}(X/Y;W)$ contains no lines of $N_1(X/Y;W)$.*

Proof. Suppose that $\overline{\mathrm{NE}}(X/Y;W)$ contains a line of $N_1(X/Y;W)$. Then we can take $\Gamma \in \overline{\mathrm{NE}}(X/Y;W)$ such that $\Gamma, -\Gamma \in \overline{\mathrm{NE}}(X/Y;W)$. We take a π-ample $\mathbb{R}$-line bundle $\mathcal{A}$ on X. By definition, $\mathcal{A}$ is π-nef over W. Therefore, we obtain $\mathcal{A} \cdot \Gamma \geq 0$ and $-\mathcal{A} \cdot \Gamma \geq 0$. This means that $\mathcal{A} \cdot \Gamma = 0$. On the other hand, after shrinking Y around W suitably, we can take a line bundle $\mathcal{M}$ on X such that $\Gamma \cdot \mathcal{M} > 0$ since $\Gamma \neq 0$ in $N_1(X/Y;W)$. Since $\mathcal{A}$ is π-ample, $m\mathcal{A} - \mathcal{M}$ is also π-ample over some open neighborhood of W, where m is a large positive integer. This implies that $(m\mathcal{A} - \mathcal{M}) \cdot \Gamma \geq 0$ since $\Gamma \in \overline{\mathrm{NE}}(X/Y;W)$. Thus, $m\mathcal{A} \cdot \Gamma \geq \mathcal{M} \cdot \Gamma > 0$ holds. Hence we obtain $\mathcal{A} \cdot \Gamma > 0$. This is a contradiction. Therefore, there are no lines in $\overline{\mathrm{NE}}(X/Y;W)$. □

The following theorem is Kleiman's ampleness criterion for projective morphisms between complex analytic spaces (see [Na1, Proposition 4.7]).

Theorem 4.4.5 (Kleiman's ampleness criterion)**.** *Let $\pi\colon X \to Y$ be a projective morphism between complex analytic spaces and let W be a compact subset of Y such that the dimension of $N^1(X/Y; W)$ is finite. Let $\mathcal{L}$ be an $\mathbb{R}$-line bundle on X. Then the following conditions are equivalent.*

(i) *$\mathcal{L}$ is π-ample over W.*

(ii) *$\mathcal{L}$ is π-ample over some open neighborhood U of W.*

(iii) *$\mathcal{L}$ is positive on $\overline{\mathrm{NE}}(X/Y; W) \setminus \{0\}$.*

Proof. We have already proved the equivalence of (i) and (ii) in Lemma 2.2.2 without assuming $\dim_{\mathbb{R}} N^1(X/Y; W) < \infty$.

Step 1. In this step, we will prove that (i) follows from (iii).

We assume that $\mathcal{L}$ is positive on $\overline{\mathrm{NE}}(X/Y; W)\setminus\{0\}$. Then we can take a π-ample $\mathbb{R}$-line bundle $\mathcal{A}$ on X such that $\mathcal{N} := \mathcal{L} - \mathcal{A}$ is non-negative on $\overline{\mathrm{NE}}(X/Y; W)$. This means that $\mathcal{N}|_{\pi^{-1}(w)}$ is nef for every $w \in W$. Since $\mathcal{L}|_{\pi^{-1}(w)} = \mathcal{N}|_{\pi^{-1}(w)} + \mathcal{A}|_{\pi^{-1}(w)}$, $\mathcal{N}|_{\pi^{-1}(w)}$ is nef, and $\mathcal{A}|_{\pi^{-1}(w)}$ is ample, $\mathcal{L}|_{\pi^{-1}(w)}$ is ample by the usual Kleiman's ampleness criterion. Hence (i) follows from (iii).

Step 2. In this step, we will prove that (iii) follows from (ii).

We assume that $\mathcal{L}$ is π-ample over some open neighborhood U of W. By replacing Y with U, we may assume that $Y = U$. In the proof of Lemma 4.4.4, we have already checked that $\mathcal{L} \cdot \Gamma > 0$ for every $\Gamma \in \overline{\mathrm{NE}}(X/Y; W) \setminus \{0\}$. This means that (iii) follows from (ii).

We finish the proof. □

From now on, we always assume that the dimension of $N^1(X/Y; W)$ is finite. In order to formulate the cone and contraction theorem, we need the following definitions.

Definition 4.4.6. Let $\pi\colon X \to Y$ be a projective morphism of complex analytic spaces and let W be a compact subset of Y. Let X be a normal complex variety and let Δ be an effective $\mathbb{R}$-divisor on X such that $K_X + \Delta$ is $\mathbb{R}$-Cartier. We assume that the dimension of $N^1(X/Y; W)$ is finite. Then we define a subcone

$$\overline{\mathrm{NE}}(X/Y; W)_{\mathrm{Nlc}(X,\Delta)}$$

of $\overline{\mathrm{NE}}(X/Y; W)$ as the closure of the convex cone spanned by the projective integral curves C on $\mathrm{Nlc}(X, \Delta)$ such that $\pi(C)$ is a point of W. Let D be an element of $N^1(X/Y; W)$. We define

$$D_{\geq 0} := \{z \in N_1(X/Y; W) \mid D \cdot z \geq 0\}.$$

Similarly, we can define $D_{>0}$, $D_{\leq 0}$, and $D_{<0}$. We also define

$$D^{\perp} := \{z \in N_1(X/Y; W) \mid D \cdot z = 0\}.$$

We use the following notation

$$\overline{\mathrm{NE}}(X/Y;W)_{D\geq 0}:=\overline{\mathrm{NE}}(X/Y;W)\cap D_{\geq 0},$$

and similarly for >0, ≤ 0, and <0.

Definition 4.4.7. An *extremal face* of the Kleiman–Mori cone $\overline{\mathrm{NE}}(X/Y;W)$ is a non-zero subcone $F\subset\overline{\mathrm{NE}}(X/Y;W)$ such that $z,z'\in\overline{\mathrm{NE}}(X/Y;W)$ and $z+z'\in F$ imply that $z,z'\in F$. Equivalently, $F=\overline{\mathrm{NE}}(X/Y;W)\cap\mathcal{H}^{\perp}$ for some $\mathbb{R}$-line bundle $\mathcal{H}$ which is defined on some open neighborhood of $\pi^{-1}(W)$ and is π-nef over W. We call $\mathcal{H}$ a *support function* of F. An *extremal ray* is a one-dimensional extremal face.

(1) An extremal face F is called $(K_X+\Delta)$*-negative* if

$$F\cap\overline{\mathrm{NE}}(X/Y;W)_{(K_X+\Delta)\geq 0}=\{0\}.$$

(2) An extremal face F is called *rational* if we can choose a $\mathbb{Q}$-line bundle $\mathcal{H}$, which is defined on some open neighborhood of $\pi^{-1}(W)$ and is π-nef over W, as a support function of F.

(3) An extremal face F is called *relatively ample at* $\mathrm{Nlc}(X,\Delta)$ if

$$F\cap\overline{\mathrm{NE}}(X/Y;W)_{\mathrm{Nlc}(X,\Delta)}=\{0\}.$$

Equivalently, $\mathcal{H}|_{\mathrm{Nlc}(X,\Delta)}$ is $\pi|_{\mathrm{Nlc}(X,\Delta)}$-ample over W for every support function $\mathcal{H}$ of F.

(4) An extremal face F is called *contractible at* $\mathrm{Nlc}(X,\Delta)$ if it has a rational support function $\mathcal{H}$ such that $\mathcal{H}|_{\mathrm{Nlc}(X,\Delta)}$ is $\pi|_{\mathrm{Nlc}(X,\Delta)}$-semiample over some open neighborhood of W.

We make a remark on (3) in Definition 4.4.7.

Remark 4.4.8. In (3) in Definition 4.4.7, the condition that F is relatively ample at $\mathrm{Nlc}(X,\Delta)$ implies that the support function $\mathcal{H}$ of F is positive on

$$\overline{\mathrm{NE}}(X/Y;W)_{\mathrm{Nlc}(X,\Delta)}\setminus\{0\}.$$

Let $\mathcal{A}$ be a π-ample $\mathbb{R}$-line bundle on X. Then $\mathcal{N}:=\mathcal{H}-\varepsilon\mathcal{A}$ is positive on $\overline{\mathrm{NE}}(X/Y;W)_{\mathrm{Nlc}(X,\Delta)}\setminus\{0\}$ for some $0<\varepsilon\ll 1$. Thus, it is easy to see that $\mathcal{N}|_{\mathrm{Nlc}(X,\Delta)}$ is $\pi|_{\mathrm{Nlc}(X,\Delta)}$-nef over W. Note that $\mathcal{A}|_{\mathrm{Nlc}(X,\Delta)}$ is obviously $\pi|_{\mathrm{Nlc}(X,\Delta)}$-ample. Hence $\mathcal{H}|_{\mathrm{Nlc}(X,\Delta)}$ is $\pi|_{\mathrm{Nlc}(X,\Delta)}$-ample over W.

4.5 Nakayama's finiteness

In this section, we quickly recall Nakayama's finiteness. As we saw in Section 4.4, for the cone and contraction theorem in this book, we need the assumption that the dimension of $N^1(X/Y;W)$ is finite. In general, the dimension of $N^1(X/Y;W)$ may be infinite. The author learned the following example from Noboru Nakayama.

Example 4.5.1 (Nakayama). Let $\pi\colon X \to Y$ be a projective surjective morphism of complex analytic spaces such that X is a normal complex variety with $\dim X \geq 2$ and that $Y = \{z \in \mathbb{C} \mid |z| < 2\}$. We put

$$W := \left\{ \frac{1}{n} \,\middle|\, n \in \mathbb{Z}_{>0} \right\} \cup \{0\}.$$

Then W is a compact subset of Y. It is obvious that W has infinitely many connected components. In this case, we can see that the abelian group $A^1(X/Y; W)$ is not finitely generated. Hence we have $\dim_{\mathbb{R}} N^1(X/Y; W) = \infty$.

The following theorem gives an important and useful sufficient condition for the finite-dimensionality of $N^1(X/Y; W)$. We state it here for the reader's convenience.

Theorem 4.5.2 (Nakayama's finiteness, see [Na2, Chapter II. 5.19. Lemma]). *Let $\pi\colon X \to Y$ be a projective surjective morphism of complex analytic spaces such that W is a compact subset of Y. We assume that $W \cap V$ has only finitely many connected components for every analytic subset V defined over an open neighborhood of W. Then $A^1(X/Y; W)$ is a finitely generated abelian group.*

Proof. For the details, see [Fu13, 4.1. Nakayama's finiteness]. □

In this book, we do not need Theorem 4.5.2 except in the proof of Corollary 1.1.9. We only need the assumption that the dimension of $N^1(X/Y; W)$ is finite.

Remark 4.5.3. Let $\pi\colon X \to Y$ be a smooth projective surjective morphism between smooth irreducible complex analytic spaces. Let W be a compact subset of Y. Assume that W is connected. Then we can easily check that the dimension of $N^1(X/Y; W)$ is finite. However, $W \cap V$ may have infinitely many connected components for some analytic subset V defined over an open neighborhood of W.

We close this section with some remarks on Nakayama's fundamental paper [Na1].

Remark 4.5.4. Example 4.5.1 shows that [Na1, Proposition 4.3] is not correct. In [Fu3, Section 4], we gave an alternative simple proof of [Na1, Theorem 5.5] (see also [Fu3, 5.3]).

4.6 Cone theorem

In this section, we will explain the cone and contraction theorem of normal pairs for projective morphisms between complex analytic spaces. The proof given in this section is essentially the same as that in [Fu4] for algebraic varieties. The main ingredients of this section is the basepoint-free theorem (see Theorem 4.2.1) and the rationality theorem (see Theorem 4.3.1).

We first treat the contraction theorem, which is a direct consequence of the basepoint-free theorem: Theorem 4.2.1. We will use it in the proof of the cone theorem: Theorem 4.6.2.

Theorem 4.6.1 (Contraction theorem). *Let $\pi\colon X \to Y$ be a projective morphism of complex analytic spaces such that X is a normal complex variety and let W be a compact subset of Y such that the dimension of $N^1(X/Y; W)$ is finite. Let Δ be an effective $\mathbb{R}$-divisor on X such that $K_X + \Delta$ is $\mathbb{R}$-Cartier. Let $\mathcal{H}$ be a line bundle which is defined on some open neighborhood of $\pi^{-1}(W)$ and is π-nef over W such that the extremal face*

$$F = \mathcal{H}^{\perp} \cap \overline{\mathrm{NE}}(X/Y; W)$$

is $(K_X+\Delta)$-negative and contractible at $\mathrm{Nlc}(X, \Delta)$. *Then, after shrinking Y around W suitably, there exists a projective morphism $\varphi_F\colon X \to Z$ over Y with the following properties.*

(1) *Let C be a projective integral curve on X such that $\pi(C)$ is a point of W. Then $\varphi_F(C)$ is a point if and only if the numerical equivalence class $[C]$ of C is in F.*
(2) *The natural map $\mathcal{O}_Z \to (\varphi_F)_*\mathcal{O}_X$ is an isomorphism.*
(3) *Let $\mathcal{L}$ be a line bundle on X such that $\mathcal{L}\cdot C = 0$ for every curve C with $[C] \in F$. Assume that $\mathcal{L}^{\otimes m}|_{\mathrm{Nlc}(X,\Delta)}$ is $\varphi_F|_{\mathrm{Nlc}(X,\Delta)}$-generated for every $m \gg 0$. Then, after shrinking Y around W suitably again, there exists a line bundle $\mathcal{L}_Z$ on Z such that $\mathcal{L} \simeq \varphi_F^*\mathcal{L}_Z$ holds.*

As we mentioned above, Theorem 4.6.1 easily follows from the basepoint-free theorem: Theorem 4.2.1.

Proof of Theorem 4.6.1. Since F is contractible at $\mathrm{Nlc}(X, \Delta)$ by assumption, we may assume that $\mathcal{H}|_{\mathrm{Nlc}(X,\Delta)}$ is $\pi|_{\mathrm{Nlc}(X,\Delta)}$-semiample over some open neighborhood of W. Since F is $(K_X+\Delta)$-negative by assumption, we can take some positive integer a such that $a\mathcal{H} - (K_X + \Delta)$ is π-ample over W. By the basepoint-free theorem (see Theorem 4.2.1), after shrinking Y around W suitably, $\mathcal{H}^{\otimes m}$ is π-generated for some positive integer m. We take the Stein factorization of the associated morphism. Then we can obtain a contraction morphism $\varphi_F\colon X \to Z$ over Y satisfying the properties (1) and (2). We consider $\varphi_F\colon X \to Z$ and $\overline{\mathrm{NE}}(X/Z; \pi_Z^{-1}(W))$, where $\pi_Z\colon Z \to Y$ is the structure morphism. Then $\overline{\mathrm{NE}}(X/Z; \pi_Z^{-1}(W)) = F$ holds by construction, $\mathcal{L}$ is numerically trivial over $\pi_Z^{-1}(W)$, and $-(K_X + \Delta)$ is φ_F-ample over $\pi_Z^{-1}(W)$. We use the basepoint-free theorem over Z (see Theorem 4.2.1). Then, after shrinking Z around $\pi_Z^{-1}(W)$ suitably, both $\mathcal{L}^{\otimes m}$ and $\mathcal{L}^{\otimes(m+1)}$ are pull-backs of line bundles on Z. Their difference gives a line bundle $\mathcal{L}_Z$ on Z such that $\mathcal{L} \simeq \varphi_F^*\mathcal{L}_Z$ holds. We finish the proof of Theorem 4.6.1. □

The following theorem is the main result of this section, which is the cone theorem of normal pairs for projective morphisms between complex analytic spaces.

Theorem 4.6.2 (Cone theorem, see [Fu4, Theorem 16.6]). *Let $\pi\colon X \to Y$ be a projective morphism of complex analytic spaces such that X is a normal complex variety and let W be a compact subset of Y such that the dimension of $N^1(X/Y; W)$ is finite. Let Δ be an effective $\mathbb{R}$-divisor on X such that $K_X+\Delta$ is $\mathbb{R}$-Cartier. Then we have the following properties.*

(1) *We can write*

$$\overline{\mathrm{NE}}(X/Y;W)=\overline{\mathrm{NE}}(X/Y;W)_{(K_X+\Delta)\geq 0}+\overline{\mathrm{NE}}(X/Y;W)_{\mathrm{Nlc}(X,\Delta)}+\sum R_j,$$

where R_j*'s are the* $(K_X+\Delta)$*-negative extremal rays of* $\overline{\mathrm{NE}}(X/Y;W)$ *that are rational and relatively ample at* $\mathrm{Nlc}(X,\Delta)$*. In particular, each* R_j *is spanned by an integral curve* C_j *on* X *such that* $\pi(C_j)$ *is a point of* W*.*

(2) *Let* $\mathcal{A}$ *be a* π*-ample* $\mathbb{R}$*-line bundle defined on some open neighborhood of* $\pi^{-1}(W)$*. Then there are only finitely many* R_j*'s included in*

$$\overline{\mathrm{NE}}(X/Y;W)_{(K_X+\Delta+\mathcal{A})<0}.$$

In particular, the R_j*'s are discrete in the half-space*

$$\overline{\mathrm{NE}}(X/Y;W)_{(K_X+\Delta)<0}.$$

(3) *Let* F *be a* $(K_X+\Delta)$*-negative extremal face of* $\overline{\mathrm{NE}}(X/Y;W)$ *that is relatively ample at* $\mathrm{Nlc}(X,\Delta)$*. Then* F *is a rational face. In particular,* F *is contractible at* $\mathrm{Nlc}(X,\Delta)$*.*

By combining Theorem 4.6.2 with Theorem 4.6.1, we obtain the cone and contraction theorem of normal pairs for projective morphisms between complex analytic spaces.

Proof of Theorem 4.6.2. Without loss of generality, we can freely shrink Y around W suitably throughout this proof. From Step 1 to Step 5, we will prove Theorem 4.6.2 under the extra assumption that $K_X+\Delta$ is $\mathbb{Q}$-Cartier. Then, in Step 6, we will treat the general case. We note that we may assume that $\dim_{\mathbb{R}} N_1(X/Y;W)\geq 2$ and $K_X+\Delta$ is not π-nef over W. Otherwise, the theorem is obvious.

Step 1. In this step, we will prove:

Claim 1. *When* $K_X+\Delta$ *is* $\mathbb{Q}$*-Cartier, the following equality*

$$\overline{\mathrm{NE}}(X/Y;W)=\overline{\overline{\mathrm{NE}}(X/Y;W)_{(K_X+\Delta)\geq 0}+\overline{\mathrm{NE}}(X/Y;W)_{\mathrm{Nlc}(X,\Delta)}+\sum_F F}$$

holds, where F*'s vary among all rational proper* $(K_X+\Delta)$*-negative extremal faces that are relatively ample at* $\mathrm{Nlc}(X,\Delta)$*.*

We note that in Claim 1 $\overline{}$ denotes the closure with respect to the real topology.

Proof of Claim 1. We put

$$\mathfrak{B}=\overline{\overline{\mathrm{NE}}(X/Y;W)_{(K_X+\Delta)\geq 0}+\overline{\mathrm{NE}}(X/Y;W)_{\mathrm{Nlc}(X,\Delta)}+\sum_F F}.$$

The inclusion $\overline{\mathrm{NE}}(X/Y;W) \supset \mathfrak{B}$ obviously holds by definition. We note that each F is spanned by curves on X mapped to points in W by Theorem 4.6.1 (1). From now on, we suppose $\overline{\mathrm{NE}}(X/Y;W) \neq \mathfrak{B}$. Then we will derive a contradiction. We can take a separating function M which is a line bundle on some open neighborhood of $\pi^{-1}(W)$ and is not a multiple of $K_X+\Delta$ in $N^1(X/Y;W)$ such that $M>0$ on $\mathfrak{B}\setminus\{0\}$ and $M\cdot z_0<0$ for some $z_0 \in \overline{\mathrm{NE}}(X/Y;W)$. Let $\mathcal{C}$ be the dual cone of $\overline{\mathrm{NE}}(X/Y;W)_{(K_X+\Delta)\geq 0}$, that is,

$$\mathcal{C}=\left\{D \in N^1(X/Y;W) \mid D\cdot z\geq 0 \text{ for } z\in \overline{\mathrm{NE}}(X/Y;W)_{(K_X+\Delta)\geq 0}\right\}.$$

Then $\mathcal{C}$ is generated by $K_X+\Delta$ and $\mathbb{R}$-line bundles on X which are π-nef over W. Since $M>0$ on $\overline{\mathrm{NE}}(X/Y;W)_{(K_X+\Delta)\geq 0}\setminus\{0\}$, M is in the interior of $\mathcal{C}$. Hence there exists a π-ample $\mathbb{R}$-line bundle A such that

$$M-A=L'+p(K_X+\Delta)$$

in $N^1(X/Y;W)$, where L' is an $\mathbb{R}$-line bundle on some open neighborhood of $\pi^{-1}(W)$ which is π-nef over W, and p is a non-negative rational number. Therefore, M is expressed in the form

$$M=H+p(K_X+\Delta)$$

in $N^1(X/Y;W)$, where $H=M-p(K_X+\Delta)$ is a $\mathbb{Q}$-line bundle on X. Note that H is π-ample over W since $H=A+L'$ in $N^1(X/Y;W)$. The rationality theorem (see Theorem 4.3.1) implies that there exists a positive rational number $r<p$ such that

$$L=H+r(K_X+\Delta)$$

is π-nef over W but not π-ample over W, and $L|_{\mathrm{Nlc}(X,\Delta)}$ is $\pi|_{\mathrm{Nlc}(X,\Delta)}$-ample over W. We note that $L\neq 0$ in $N^1(X/Y;W)$ since M is not a multiple of $K_X+\Delta$. Thus the extremal face F_L associated to the support function L is contained in $\mathfrak{B}$, which implies $M>0$ on F_L. Therefore, $p<r$. It is a contradiction. This completes the proof of Claim 1. □

Step 2. In this step, we will prove:

Claim 2. *In the equality in Claim 1, we can assume that every extremal face F is one-dimensional.*

Proof of Claim 2. Let F be a rational proper $(K_X+\Delta)$-negative extremal face that is relatively ample at $\mathrm{Nlc}(X,\Delta)$. We assume that $\dim F\geq 2$. After shrinking Y around W suitably, we can take the contraction morphism $\varphi_F: X\to Z$ over Y associated to F (see Theorem 4.6.1). We note that $F=\overline{\mathrm{NE}}(X/Z;\pi_Z^{-1}(W))$, where $\pi_Z\colon Z\to Y$ is the structure morphism, and that $-(K_X+\Delta)$ is φ_F-ample over $\pi_Z^{-1}(W)$ by construction. By Claim 1 in Step 1, we obtain

$$F=\overline{\mathrm{NE}}(X/Z;\pi_Z^{-1}(W))=\overline{\sum_G G}, \tag{4.6.1}$$

where the G's in (4.6.1) are the rational proper $(K_X+\Delta)$-negative extremal faces of $\overline{\mathrm{NE}}(X/Z;\pi_Z^{-1}(W))$. We note that $\overline{\mathrm{NE}}(X/Z;\pi_Z^{-1}(W))_{\mathrm{Nlc}(X,\Delta)}=0$ holds because φ_F embeds $\mathrm{Nlc}(X,\Delta)$ into Z. The G's are also $(K_X+\Delta)$-negative extremal faces of $\overline{\mathrm{NE}}(X/Y;W)$ that are ample at $\mathrm{Nlc}(X,\Delta)$ with $\dim G<\dim F$. By induction, we finally obtain

$$\overline{\mathrm{NE}}(X/Y;W)=\overline{\overline{\mathrm{NE}}(X/Y;W)_{(K_X+\Delta)\geq 0}+\overline{\mathrm{NE}}(X/Y;W)_{\mathrm{Nlc}(X,\Delta)}+\sum R_j}, \quad (4.6.2)$$

where the R_j's are $(K_X+\Delta)$-negative rational extremal rays. Note that each R_j does not intersect $\overline{\mathrm{NE}}(X/Y;W)_{\mathrm{Nlc}(X,\Delta)}$. We finish the proof of Claim 2. □

Step 3. In this step, we still assume that $K_X+\Delta$ is $\mathbb{Q}$-Cartier. We will finish the proof of (1) when $K_X+\Delta$ is $\mathbb{Q}$-Cartier.

The contraction theorem (see Theorem 4.6.1) guarantees that for each extremal ray R_j, which is $(K_X+\Delta)$-negative, rational, and relatively ample at $\mathrm{Nlc}(X,\Delta)$, there exists a projective integral curve C_j on X such that $[C_j]\in R_j$. Let $\psi_j:X\to Z_j$ be the contraction morphism of R_j over Y after shrinking Y around W suitably, and let A be a π-ample line bundle on X. Let $\pi_{Z_j}:Z_j\to Y$ be the structure morphism. We set

$$r_j=-\frac{A\cdot C_j}{(K_X+\Delta)\cdot C_j}.$$

Then $A+r_j(K_X+\Delta)$ is ψ_j-nef over $\pi_{Z_j}^{-1}(W)$ but not ψ_j-ample over $\pi_{Z_j}^{-1}(W)$, and

$$(A+r_j(K_X+\Delta))|_{\mathrm{Nlc}(X,\Delta)}$$

is $\psi_j|_{\mathrm{Nlc}(X,\Delta)}$-ample over $\pi_{Z_j}^{-1}(W)$. By the rationality theorem (see Theorem 4.3.1), expressing $r_j=u_j/v_j$ with $u_j,v_j\in\mathbb{Z}_{>0}$ and $(u_j,v_j)=1$, we have the inequality $v_j\leq a(\dim X+1)$, where a is a positive integer such that $a(K_X+\Delta)$ is Cartier. After shrinking Y around W suitably, we take π-ample line bundles $H_1,H_2,\ldots,H_{\rho-1}$ on X such that $K_X+\Delta$ and the H_i's form a basis of $N^1(X/Y;W)$, where $\rho=\dim_{\mathbb{R}}N^1(X/Y;W)<\infty$. As we saw above, the intersection of the extremal rays R_j with the hyperplane

$$\{z\in N_1(X/Y;W)\mid a(K_X+\Delta)\cdot z=-1\}$$

in $N_1(X/Y;W)$ lie on the lattice

$$\Lambda=\left\{z\in N_1(X/Y;W)\mid a(K_X+\Delta)\cdot z=-1, H_i\cdot z\in(a(a(\dim X+1))!)^{-1}\mathbb{Z}\right\}.$$

This implies that the extremal rays are discrete in the half space

$$\{z\in N_1(X/Y;W)\mid (K_X+\Delta)\cdot z<0\}.$$

Thus we can omit the closure sign $\overline{}$ from the formula (4.6.2) and this completes the proof of (1) when $K_X+\Delta$ is $\mathbb{Q}$-Cartier.

Step 4. In this step, we will prove (2) under the assumption that $K_X + \Delta$ is $\mathbb{Q}$-Cartier.

Let $\mathcal{A}$ be a π-ample $\mathbb{R}$-line bundle on X. We choose $0 < \varepsilon_i \ll 1$ for $1 \leq i \leq \rho-1$ such that $\mathcal{A}-\sum_{i=1}^{\rho-1} \varepsilon_i H_i$ is still π-ample. Then the R_j's included in $(K_X+\Delta+\mathcal{A})_{<0}$ correspond to some elements of the above lattice Λ in Step 3 for which

$$\sum_{i=1}^{\rho-1} \varepsilon_i H_i \cdot z < 1/a.$$

Therefore, we obtain (2) when $K_X + \Delta$ is $\mathbb{Q}$-Cartier.

Step 5. In this step, we will prove (3) under the extra assumption that $K_X + \Delta$ is $\mathbb{Q}$-Cartier.

Let F be a $(K_X + \Delta)$-negative extremal face as in (3). The vector space $V = F^{\perp} \subset N^1(X/Y; W)$ is defined over $\mathbb{Q}$ because F is generated by some of the R_j's. There exists a π-ample $\mathbb{R}$-line bundle $\mathcal{A}$ such that F is contained in $(K_X+\Delta+\mathcal{A})_{<0}$. Let $\langle F\rangle$ be the vector space spanned by F. We put

$$\mathcal{C}_F := \overline{\mathrm{NE}}(X/Y; W)_{(K_X+\Delta+\mathcal{A})\geq 0} + \overline{\mathrm{NE}}(X/Y; W)_{\mathrm{Nlc}(X,\Delta)} + \sum_{R_j \not\subset F} R_j.$$

Then $\mathcal{C}_F$ is a closed cone,

$$\overline{\mathrm{NE}}(X/Y; W) = \mathcal{C}_F + F,$$

and

$$\mathcal{C}_F \cap \langle F\rangle = \{0\}.$$

The support functions of F are the elements of V that are positive on $\mathcal{C}_F \setminus \{0\}$. This is a non-empty open subset of V and thus it contains a rational element that, after scaling, gives a line bundle L defined over some open neighborhood of W such that L is π-nef over W and that $F = L^{\perp} \cap \overline{\mathrm{NE}}(X/Y; W)$. Therefore, F is rational. Hence, we obtain (3) when $K_X + \Delta$ is $\mathbb{Q}$-Cartier.

We finish the proof of Theorem 4.6.2 under the extra assumption that $K_X+\Delta$ is $\mathbb{Q}$-Cartier. Therefore, from now on, we can freely use Theorem 4.6.2 when $K_X + \Delta$ is $\mathbb{Q}$-Cartier.

Step 6. In this final step, we will treat the general case. This means that we will treat the case where $K_X + \Delta$ is $\mathbb{R}$-Cartier.

Let $\mathcal{A}$ be a π-ample $\mathbb{R}$-line bundle on X. First we will prove (2). By Lemma 2.3.5, after shrinking Y around W suitably, we can take effective $\mathbb{Q}$-divisors

$$\Delta_1, \ldots, \Delta_k$$

on X and positive real numbers $r_1, \ldots, r_k$ with $\sum_{i=1}^{k} r_i = 1$ such that

$$K_X + \Delta = \sum_{i=1}^{k} r_i(K_X + \Delta_i)$$

and that $\mathcal{J}_{\mathrm{NLC}}(X,\Delta_i)=\mathcal{J}_{\mathrm{NLC}}(X,\Delta)$ holds for every i. Since $K_X+\Delta_i$ is $\mathbb{Q}$-Cartier, there are only finitely many $(K_X+\Delta_i+\mathcal{A})$-negarive extremal rays of $\overline{\mathrm{NE}}(X/Y;W)$ which are rational and relatively ample at $\mathrm{Nlc}(X,\Delta_i)=\mathrm{Nlc}(X,\Delta)$ for every i. Therefore, since

$$K_X+\Delta+\mathcal{A}=\sum_{i=1}^{k} r_i(K_X+\Delta_i+\mathcal{A})$$

holds, there exist only finitely many $(K_X+\Delta+\mathcal{A})$-negative extremal rays of $\overline{\mathrm{NE}}(X/Y;W)$ which are rational and relatively ample at $\mathrm{Nlc}(X,\Delta)$. Thus we obtain (2) in full generality. The statement (1) is a direct and formal consequence of (2). For the details, see, for example, the proof of [Kol, Chapter III. 1.2 Theorem]. Finally, we will prove (3). Let F be a $(K_X+\Delta)$-negative extremal face of $\overline{\mathrm{NE}}(X/Y;W)$ as in (3). By using Lemma 2.3.5, after shrinking Y around W suitably, we can take an effective $\mathbb{Q}$-divisor $\Delta^\dagger$ on X, which is sufficiently close to Δ, such that $K_X+\Delta^\dagger$ is $\mathbb{Q}$-Cartier, $\mathcal{J}_{\mathrm{NLC}}(X,\Delta^\dagger)=\mathcal{J}_{\mathrm{NLC}}(X,\Delta)$ holds, and F is $(K_X+\Delta^\dagger)$-negative. Therefore, we see that F is a rational face of $\overline{\mathrm{NE}}(X/Y;W)$ by Step 5. This is what we wanted.

We finish the proof of the cone theorem. □

4.7 Proof of Theorem 1.1.8 and Corollary 1.1.9

In this section, we will prove Theorem 1.1.8 as an application of the vanishing theorem for projective quasi-log schemes (see [Fu6, Theorem 6.3.5 (ii)]). Note that Corollary 1.1.9 is an easy consequence of Theorem 1.1.8. For the details of the framework of quasi-log schemes, see [Fu6, Chapter 6], [Fu7], [Fu11], and [Fu12]. Let us start with an easy lemma.

Lemma 4.7.1 (see [FM1, Lemma 4.2]). *Let $[V,\omega]$ be a positive-dimensional irreducible projective quasi-log scheme with $\dim \mathrm{Nqlc}(V,\omega)=0$ or $\mathrm{Nqlc}(V,\omega)=\emptyset$, and let $\mathcal{M}$ be an ample line bundle on V. Assume that $\omega+r\mathcal{M}$ is numerically trivial for some real number r. Then $r\le \dim V+1$ holds.*

Proof. If $r\le 0$, then $r\le \dim V+1$ is obvious. Hence we may assume that r is positive. We consider the following short exact sequence:

$$0\to \mathcal{I}_{\mathrm{Nqlc}(V,\omega)}\to \mathcal{O}_V\to \mathcal{O}_{\mathrm{Nqlc}(V,\omega)}\to 0, \tag{4.7.1}$$

where $\mathcal{I}_{\mathrm{Nqlc}(V,\omega)}$ is the defining ideal sheaf of $\mathrm{Nqlc}(V,\omega)$ on V. Since $l\mathcal{M}-\omega$ is ample for $l>-r$, we have

$$H^i(V,\mathcal{I}_{\mathrm{Nqlc}(V,\omega)}\otimes\mathcal{M}^{\otimes l})=0$$

for every $i\neq 0$ and $l>-r$ by the vanishing theorem for quasi-log schemes (see [Fu6, Theorem 6.3.5 (ii)]). Since $\dim \mathrm{Nqlc}(V,\omega)=0$ or $\mathrm{Nqlc}(V,\omega)=\emptyset$,

$$H^i(V,\mathcal{O}_{\mathrm{Nqlc}(V,\omega)}\otimes\mathcal{M}^{\otimes l})=0$$

for every $i \neq 0$ and every l. Therefore, we obtain $H^i(V, \mathcal{M}^{\otimes l}) = 0$ for every $i \neq 0$ and $l > -r$ by (4.7.1). Let V' be the unique maximal (with respect to the inclusion) qlc stratum of $[V, \omega]$. Then we have the following short exact sequence:

$$0 \longrightarrow \operatorname{Ker} \alpha \longrightarrow \mathcal{O}_V \xrightarrow{\alpha} \mathcal{O}_{V'} \longrightarrow 0$$

such that $\dim \operatorname{Supp} \operatorname{Ker} \alpha \leq 0$ since $\dim \operatorname{Nqlc}(V, \omega) = 0$ or $\operatorname{Nqlc}(V, \omega) = \emptyset$. Hence we have $H^i(V', \mathcal{M}^{\otimes l}|_{V'}) = 0$ for every $i \neq 0$ and $l > -r$. Since $\dim V' = \dim V > 0$, it is obvious that $H^0(V', \mathcal{M}^{\otimes l}|_{V'}) = 0$ holds for every $l < 0$. We consider

$$\chi(t) := \sum_{i=0}^{\dim V'} (-1)^i \dim_{\mathbb{C}} H^i(V', \mathcal{M}^{\otimes t}|_{V'}).$$

Then it is well known that $\chi(t)$ is a non-trivial polynomial of $\deg \chi(t) = \dim V' = \dim V$. By the above observation, $\chi(l) = 0$ for $l \in \mathbb{Z}$ with $-r < l < 0$. This implies that $r \leq \dim V + 1$. We finish the proof. $\square$

Let us start the proof of Theorem 1.1.8.

Proof of Theorem 1.1.8. We put $\mathcal{L} := f^*\mathcal{A}_{Y^\flat}$. Without loss of generality, we may assume that $\dim X \geq 1$. Since $\mathcal{L}$ is π-nef over W, R is a $(K_X + \Delta)$-negative extremal ray of $\overline{\mathrm{NE}}(X/Y; W)$. Therefore, by Theorem 4.6.2 (3) and Theorem 4.6.1, after shrinking Y around W suitably, we obtain a contraction morphism $\varphi_R \colon X \to Z$ over Y associated to R. It is sufficient to prove that $R \cdot \mathcal{L} = 0$ holds.

Step 1. In this step, we will treat the case where $\dim Z = 0$.

From now on, we assume that $\dim Z = 0$ holds. Then X is projective with $\rho(X) = 1$ and $\mathcal{L}$ is a nef line bundle on X in the usual sense. Suppose that $\mathcal{L}$ is ample. Then we obtain that $K_X + \Delta + r\mathcal{L}$ is numerically trivial for some $r > \dim X + 1$ since $(K_X + \Delta + (\dim X + 1)\mathcal{L}) \cdot R < 0$ and $\rho(X) = 1$. We note that $[X, K_X + \Delta]$ naturally becomes an irreducible projective quasi-log scheme with $\operatorname{Nqlc}(X, K_X + \Delta) = \emptyset$ (see, for example, [Fu6, 6.4.1]). Therefore, we get a contradiction by Lemma 4.7.1. This implies that $\mathcal{L}$ is numerically trivial, that is, $R \cdot f^*\mathcal{A}_{Y^\flat} = R \cdot \mathcal{L} = 0$. This is what we wanted.

Step 2. In this step, we will treat the case where $\dim Z \geq 1$.

From now on, we assume that $\dim Z \geq 1$ holds. Then we can always take a point $P \in Z$ such that $\dim \varphi_R^{-1}(P) \geq 1$. We shrink Z around P and assume that Z is Stein. Then we can take an effective $\mathbb{R}$-Cartier divisor B on Z such that $(X, \Delta + \varphi_R^* B)$ is log canonical outside $\varphi_R^{-1}(P)$, there exists a positive-dimensional log canonical center C of $(X, \Delta + \varphi_R^* B)$ with $\varphi_R(C) = P$, and $\dim \operatorname{Nlc}(X, \Delta + \varphi_R^* B) = 0$ or $\operatorname{Nlc}(X, \Delta + \varphi_R^* B) = \emptyset$. After shrinking Z around P suitably again, we can take a projective bimeromorphic morphism $f \colon V \to X$ from a smooth complex variety V such that $f^{-1}(C)$ and the exceptional locus $\operatorname{Exc}(f)$ of f are both simple normal crossing divisors on V and that the union of $f^{-1}(C)$, $\operatorname{Exc}(f)$, and

$\operatorname{Supp}\left(f_*^{-1}(\Delta+\varphi_R^*B)\right)$ is a simple normal crossing divisor on V (see [BieM2, Theorem 13.2]). We define B_V by the formula $K_V+B_V=f^*(K_X+\Delta+\varphi^*B)$. Then we see that $\operatorname{Supp} B_V$ is a simple normal crossing divisor on V. By shrinking X around C, we assume that $(X\setminus C)\cap \operatorname{Nlc}(X,\Delta+\varphi_R^*B)=\emptyset$. Let T be the union of the irreducible components of $B_V^{=1}$ that are mapped to C by f. We define B_T by adjunction: $K_T+B_T=(K_V+B_V)|_T$. We consider the following short exact sequence:

$$
\begin{aligned}
0 &\to \mathcal{O}_V(\lceil -(B_V^{<1})\rceil - \lfloor B_V^{>1}\rfloor - T) \to \mathcal{O}_V(\lceil -(B_V^{<1})\rceil - \lfloor B_V^{>1}\rfloor)\\
&\to \mathcal{O}_T(\lceil -(B_T^{<1})\rceil - \lfloor B_T^{>1}\rfloor) \to 0.
\end{aligned}
$$

We note that

$$
\lceil -(B_V^{<1})\rceil - \lfloor B_V^{>1}\rfloor - T-(K_V+\{B_V\}+B_V^{=1}-T) = -f^*(K_X+\Delta+\varphi^*B). \tag{4.7.2}
$$

By taking R^if_*, we have a long exact sequence:

$$
\begin{aligned}
0 &\longrightarrow f_*\mathcal{O}_V(\lceil -(B_V^{<1})\rceil - \lfloor B_V^{>1}\rfloor - T)\\
&\longrightarrow \mathcal{J}_{\mathrm{NLC}}(X,\Delta+\varphi^*B)\\
&\longrightarrow f_*\mathcal{O}_T(\lceil -(B_T^{<1})\rceil - \lfloor B_T^{>1}\rfloor)\\
&\xrightarrow{\ \delta\ } R^1f_*\mathcal{O}_V(\lceil -(B_V^{<1})\rceil - \lfloor B_V^{>1}\rfloor - T) \longrightarrow \cdots.
\end{aligned} \tag{4.7.3}
$$

The support of $f_*\mathcal{O}_T(\lceil -(B_T^{<1})\rceil - \lfloor B_T^{>1}\rfloor)$ is contained in C since $f(T)=C$. On the other hand, any associated subvarieties of $R^1f_*\mathcal{O}_V(\lceil -(B_V^{<1})\rceil - \lfloor B_V^{>1}\rfloor - T)$ are not contained in C by (4.7.2) and Theorem 3.1.5 (1). Hence, the connecting homomorphism δ in (4.7.3) is zero. We put

$$
\mathcal{J}:=f_*\mathcal{O}_V(\lceil -(B_V^{<1})\rceil - \lfloor B_V^{>1}\rfloor - T).
$$

Then it is an ideal sheaf contained in $\mathcal{J}_{\mathrm{NLC}}(X,\Delta+\varphi^*B)$. Let X' denote the closed analytic subspace of X defined by $\mathcal{J}$. By applying the snake lemma to the following commutative diagram:

$$
\begin{array}{ccccccccc}
0 & \longrightarrow & \mathcal{J} & \longrightarrow & \mathcal{J}_{\mathrm{NLC}}(X,\Delta+\varphi^*B) & \longrightarrow & f_*\mathcal{O}_T(\lceil -(B_T^{<1})\rceil - \lfloor B_T^{>1}\rfloor) & \longrightarrow & 0\\
 & & \downarrow & & \downarrow & & \downarrow & & \\
0 & \longrightarrow & \mathcal{O}_X & = & \mathcal{O}_X & \longrightarrow & 0 & \longrightarrow & 0,
\end{array}
$$

we obtain the short exact sequence:

$$
0\to f_*\mathcal{O}_T(\lceil -(B_T^{<1})\rceil - \lfloor B_T^{>1}\rfloor)\to \mathcal{O}_{X'}\to \mathcal{O}_{\mathrm{Nlc}(X,\Delta+\varphi^*B)}\to 0.
$$

Since C is projective and T is projective over C by construction,

$$
(X',(K_X+\Delta)|_{X'}, f\colon (T,B_T)\to X')
$$

is a projective quasi-log scheme with $\operatorname{Nqlc}(X',(K_X+\Delta)|_{X'})=\operatorname{Nlc}(X,\Delta+\varphi_R^*B)$ by [Fu7, Theorem 4.9]. In particular, we have $\dim \operatorname{Nqlc}(X',(K_X+\Delta)|_{X'})=0$

or $\mathrm{Nqlc}(X', (K_X + \Delta)|_{X'}) = \emptyset$. We note that $X' = C$ holds set theoretically by construction. We put $\omega := (K_X + \Delta)|_{X'}$. Then $-\omega$ is ample since $\varphi_R(X') = P$.

Suppose that $R \cdot \mathcal{L} > 0$ holds. Then $\mathcal{L}' := \mathcal{L}|_{X'}$ is ample and $\omega + r\mathcal{L}'$ is numerically trivial on X' for some positive real number r with $r > \dim X + 1 > \dim X' + 1$. This is a contradiction by Lemma 4.7.1. Hence we obtain $R \cdot \mathcal{L} = 0$. Therefore, R is a $(K_X + \Delta)$-negative extremal ray of $\overline{\mathrm{NE}}(X/Y^\flat; g^{-1}(W))$.

We finish the proof. □

Remark 4.7.2. We used Theorem 3.1.5 (1) in the proof of Theorem 1.1.8. We note that Theorem 3.1.5 (1) for varieties is sufficient since V in Step 2 in the proof of Theorem 1.1.8 is irreducible.

Proof of Corollary 1.1.9. Let $P \in Y$ be any point. We put $W := \{P\}$. Then the dimension of $N^1(X/Y; W)$ is finite by Theorem 4.5.2. Suppose that $K_X + \Delta + (\dim X + 1)\mathcal{A}$ is not π-nef over W. Then there exists a $(K_X + \Delta + (\dim X + 1)\mathcal{A})$-negative extremal ray R of $\overline{\mathrm{NE}}(X/Y; W)$. We put $Y^\flat := Y$, $\mathcal{A}_{Y^\flat} := \mathcal{A}$, and $f := id_Y$. Then we use Theorem 1.1.8. Thus we obtain $R \cdot \mathcal{A} = 0$. This is a contradiction since $\mathcal{A}$ is π-ample over W. Therefore, $K_X + \Delta + (\dim X + 1)\mathcal{A}$ is π-nef over W. Since P is any point of Y, this means that $K_X + \Delta + (\dim X + 1)\mathcal{A}$ is nef over Y. □

We close this section with a remark on Theorem 1.1.8 and Corollary 1.1.9.

Remark 4.7.3. In Theorem 1.1.8 and Corollary 1.1.9, we can replace $(\dim X + 1)$ with $\dim X$ when $\pi(X)$ is not a point. We can check it easily by the proof of Theorem 1.1.8.

Chapter 5
Supplementary results

In this chapter, we will discuss some supplementary results. In Section 5.1, we will discuss lengths of extremal rational curves. As an application, in Section 5.2, we will treat Shokurov's polytopes in the complex analytic setting. The result in this section seems to be indispensable for the minimal model program with scaling in the complex analytic setting. In Section 5.3, we will prove the basepoint-free theorem for $\mathbb{R}$-divisors as an application of the cone theorem. In Section 5.4, we will discuss strong $\mathbb{Q}$-factoriality. It is indispensable for the minimal model program of complex analytic spaces. This is because a divisorial sheaf on a normal complex variety does not necessarily come from a Weil divisor. In Section 5.5, we will finish the proof of the main theorem of this book, that is, Theorem 1.1.6.

5.1 Lengths of extremal rational curves

In this section, we will quickly explain that every extremal ray is spanned by a rational curve. Our result in this section generalizes Kawamata's famous result in [Ka]. We first prove the following theorem as an application of [Fu11, Theorem 1.12].

Theorem 5.1.1. *Let $\varphi\colon X \to Z$ be a projective morphism of complex analytic spaces such that X is a normal complex variety and let Δ be an effective $\mathbb{R}$-divisor on X such that $K_X + \Delta$ is $\mathbb{R}$-Cartier. Assume that $-(K_X + \Delta)$ is φ-ample. Let P be an arbitrary point of Z. Let E be any positive-dimensional irreducible component of $\varphi^{-1}(P)$ such that $E \not\subset \operatorname{Nlc}(X, \Delta)$. Then E is covered by possibly singular rational curves ℓ with*

$$0 < -(K_X + \Delta) \cdot \ell \leq 2 \dim E.$$

In particular, E is uniruled.

In the proof of Theorem 5.1.1, we will use the theory of quasi-log schemes (see [Fu6, Chapter 6], [Fu7], [Fu11], and [Fu12]).

Proof of Theorem 5.1.1. If $\varphi(X) = P$, then $E = X$ obviously holds. In this case, the statement follows from [Fu11, Theorem 1.12] since we can see $[X, K_X + \Delta]$ as a projective quasi-log scheme (see, for example, [Fu6, 6.4.1]). Therefore, from now on, we may assume that $\varphi(X) \neq P$. We shrink Z around P and may assume that Z is Stein. Then we can take an effective $\mathbb{R}$-Cartier divisor B on Z such that E is a log canonical center of $(X, \Delta + \varphi^*B)$. After shrinking Z around P suitably again, we can take a projective bimeromorphic morphism $f\colon Y \to X$ from a smooth complex variety Y such that $f^{-1}(E)$ is a simple normal crossing divisor on Y,

$$K_Y + B_Y := f^*(K_X + \Delta + \varphi^*B),$$

and $\operatorname{Supp} B_Y$ is a simple normal crossing divisor on Y (see [BieM2, Theorem 13.2]). We may further assume that the support of the union of $f^{-1}(E)$ and $\operatorname{Supp} B_Y$ is also a simple normal crossing divisor on Y. Let T be the union of the irreducible components of $B_Y^{=1}$ that are mapped to E by f. We put $A := \lceil -(B_Y^{<1}) \rceil$ and $N := \lfloor B_Y^{>1} \rfloor$ and consider the following short exact sequence:

$$0 \to \mathcal{O}_Y(A-N-T) \to \mathcal{O}_Y(A-N) \to \mathcal{O}_T(A-N) \to 0.$$

We note that

$$A - N - T - (K_Y + \{B_Y\} + B_Y^{=1} - T) = -f^*(K_X + \Delta + \varphi^*B). \tag{5.1.1}$$

By taking $R^i f_*$, we have a long exact sequence:

$$\begin{aligned} 0 &\longrightarrow f_*\mathcal{O}_Y(A-N-T) \longrightarrow \mathcal{J}_{\mathrm{NLC}}(X, \Delta + \varphi^*B) \longrightarrow f_*\mathcal{O}_T(A-N) \\ &\overset{\delta}{\longrightarrow} R^1 f_*\mathcal{O}_Y(A-N-T) \longrightarrow \cdots . \end{aligned} \tag{5.1.2}$$

The support of $f_*\mathcal{O}_T(A-N)$ is contained in E since $f(T) = E$. On the other hand, any associated subvarieties of $R^1 f_*\mathcal{O}_Y(A-N-T)$ are not contained in E by (5.1.1) and Theorem 3.1.5 (1). Hence, the connecting homomorphism δ in (5.1.2) is zero. We put $\mathcal{J} := f_*\mathcal{O}_Y(A-N-T)$. Then it is an ideal sheaf contained in $\mathcal{J}_{\mathrm{NLC}}(X, \Delta+\varphi^*B)$. Let X' denote the closed analytic subspace of X defined by $\mathcal{J}$. Thus we obtain the following big commutative diagram.

$$\begin{array}{ccccccccc}
 & & 0 & & 0 & & & & \\
 & & \downarrow & & \downarrow & & & & \\
0 & \longrightarrow & \mathcal{J} & \longrightarrow & \mathcal{J}_{\mathrm{NLC}}(X, \Delta + \varphi^*B) & \longrightarrow & f_*\mathcal{O}_T(A-N) & \longrightarrow & 0 \\
 & & \downarrow & & \downarrow & & & & \\
 & & \mathcal{O}_X & = & \mathcal{O}_X & & & & \\
 & & \downarrow & & \downarrow & & & & \\
0 \longrightarrow f_*\mathcal{O}_T(A-N) & \longrightarrow & \mathcal{O}_{X'} & \longrightarrow & \mathcal{O}_{\mathrm{Nlc}(X, \Delta+\varphi^*B)} & \longrightarrow & 0 & & \\
 & & \downarrow & & \downarrow & & & & \\
 & & 0 & & 0 & & & &
\end{array}$$

We note that $X' = E \cup \mathrm{Nlc}(X, \Delta + \varphi^*B)$ holds set theoretically. On X', by the above big commutative diagram, we see that $\mathrm{Nlc}(X, \Delta+\varphi^*B)$ is defined by the ideal sheaf $f_*\mathcal{O}_T(A-N)$. We write $T = T' + T''$, where T'' is the union of the irreducible components of T mapped to $\mathrm{Nlc}(X, \Delta+\varphi^*B)$ by f. We put $\mathcal{I} := f_*\mathcal{O}_{T''}(A-N-T')$. Then

$$\mathcal{I} = f_*\mathcal{O}_{T''}(A-N-T') \subset f_*\mathcal{O}_T(A-N) \subset \mathcal{O}_{X'}.$$

Since $\mathcal{I} \subset f_*\mathcal{O}_T(A-N)$, $\mathcal{I}$ is zero when it is restricted to $\mathrm{Nlc}(X, \Delta + \varphi^*B)$. Since $f(T'') \subset \mathrm{Nlc}(X, \Delta + \varphi^*B)$, $\mathcal{I}$ is zero on $X' \setminus \mathrm{Nlc}(X, \Delta + \varphi^*B)$. Thus, we have

$\mathcal{I} = \{0\}$. By construction, we see that E is a closed analytic subvariety of X'. Let $\mathcal{I}_E$ be the defining ideal sheaf of E on X'. Then we obtain

$$\mathcal{I}_E \cap f_*\mathcal{O}_T(A - N) \subset f_*\mathcal{O}_{T''}(A - N - T') = \mathcal{I} = \{0\}.$$

Hence, we can see $f_*\mathcal{O}_T(A - N)$ as an ideal sheaf on E. Since E is projective and T is projective over E by construction,

$$(E, (K_X + \Delta)|_E, f\colon (T, B_T) \to E)$$

is a projective quasi-log scheme, where $K_T + B_T := (K_Y + B_Y)|_T$, by [Fu7, Theorem 4.9]. Since $\varphi(E) = P$, $-(K_X + \Delta)|_E$ is ample. Thus, by [Fu11, Theorem 1.12], E is covered by possibly singular rational curves ℓ with $0 < -(K_X + \Delta) \cdot \ell \leq 2 \dim E$. In particular, this implies that E is uniruled. □

Remark 5.1.2. We used Theorem 3.1.5 (1) in the proof of Theorem 5.1.1. We note that Theorem 3.1.5 (1) for varieties is sufficient since Y in the proof of Theorem 5.1.1 is irreducible.

Theorem 5.1.3 is an easy consequence of Theorem 5.1.1. It seems to be indispensable for the minimal model program with scaling.

Theorem 5.1.3 (Lengths of extremal rational curves). *Let $\pi\colon X \to Y$ be a projective morphism of complex analytic spaces such that X is a normal complex variety and let W be a compact subset of Y such that the dimension of $N^1(X/Y; W)$ is finite. Let Δ be an effective $\mathbb{R}$-divisor on X such that $K_X + \Delta$ is $\mathbb{R}$-Cartier. If R is a $(K_X + \Delta)$-negative extremal ray of $\overline{\mathrm{NE}}(X/Y; W)$ which is relatively ample at* $\mathrm{Nlc}(X, \Delta)$*, then there exists a rational curve ℓ spanning R with*

$$0 < -(K_X + \Delta) \cdot \ell \leq 2 \dim X.$$

Proof. By the cone and contraction theorem (see Theorems 4.6.1 and 4.6.2), after shrinking Y around W suitably, we obtain a contraction morphism $\varphi\colon X \to Z$ over Y associated to R. We note that $-(K_X + \Delta)$ is φ-ample and $\varphi\colon \mathrm{Nlc}(X, \Delta) \to \varphi(\mathrm{Nlc}(X, \Delta))$ is finite by construction. Therefore, we can find a rational curve ℓ in a fiber of φ with $0 < -(K_X + \Delta) \cdot \ell \leq 2 \dim X$ by Theorem 5.1.1. This ℓ is a desired rational curve spanning R. □

5.2 On Shokurov's polytopes

In this section, we will discuss Shokurov's polytopes for projective morphisms of complex analytic spaces. Here, we will follow the presentation in [Fu6, Section 4.7]. Let us recall the definition of *extremal curves*.

Definition 5.2.1 (Extremal curves). Let $\pi\colon X \to Y$ be a projective morphism of complex analytic spaces such that X is a normal complex variety, and let W be a compact subset of Y such that the dimension of $N^1(X/Y; W)$ is finite. A curve Γ on X is called *extremal over* W if the following properties hold.

(i) Γ generates an extremal ray R of $\overline{\mathrm{NE}}(X/Y;W)$.

(ii) There exists a π-ample line bundle $\mathcal{H}$ over some open neighborhood of W such that
$$\mathcal{H}\cdot\Gamma=\min_{\ell}\{\mathcal{H}\cdot\ell\},$$
where ℓ ranges over curves generating R.

By Lemma 5.1.3, we have:

Lemma 5.2.2. *Let $\pi\colon X\to Y$ be a projective morphism of complex analytic spaces and let (X,Δ) be a log canonical pair. Let W be a compact subset of Y such that the dimension of $N^1(X/Y;W)$ is finite. Let R be a $(K_X+\Delta)$-negative extremal ray of $\overline{\mathrm{NE}}(X/Y;W)$. If Γ is an extremal curve over W generating R, then*
$$0<-(K_X+\Delta)\cdot\Gamma\leq 2\dim X$$
holds.

Proof. By Theorem 5.1.3, we can take a rational curve ℓ spanning R such that
$$0<-(K_X+\Delta)\cdot\ell\leq 2\dim X.$$
Let $\mathcal{H}$ be a line bundle as in Definition 5.2.1. Then
$$\frac{-(K_X+\Delta)\cdot\Gamma}{\mathcal{H}\cdot\Gamma}=\frac{-(K_X+\Delta)\cdot\ell}{\mathcal{H}\cdot\ell}.$$
holds. Hence we obtain
$$0<-(K_X+\Delta)\cdot\Gamma=(-(K_X+\Delta)\cdot\ell)\cdot\frac{\mathcal{H}\cdot\Gamma}{\mathcal{H}\cdot\ell}\leq 2\dim X.$$
This is what we wanted. □

One of the main purposes of this section is to explain the following theorem, which is very well known and has already played an important role when $\pi\colon X\to Y$ is algebraic.

Theorem 5.2.3. *Let $\pi\colon X\to Y$ be a projective morphism of complex analytic spaces such that X is a normal complex variety, and let W be a compact subset of Y such that the dimension of $N^1(X/Y;W)$ is finite. Let V be a finite-dimensional affine subspace of $\mathrm{WDiv}_{\mathbb{R}}(X)$, which is defined over the rationals. We fix an $\mathbb{R}$-divisor $\Delta\in\mathcal{L}(V;\pi^{-1}(W))$, that is, $\Delta\in V$ and (X,Δ) is log canonical at $\pi^{-1}(W)$. Then we can find positive real numbers α and δ, which depend on (X,Δ) and V, with the following properties.*

(1) *If Γ is any extremal curve over W and $(K_X+\Delta)\cdot\Gamma>0$, then $(K_X+\Delta)\cdot\Gamma>\alpha$.*

(2) *If $D\in\mathcal{L}(V;\pi^{-1}(W))$, $\|D-\Delta\|<\delta$, and $(K_X+D)\cdot\Gamma\leq 0$ for an extremal curve Γ over W, then $(K_X+\Delta)\cdot\Gamma\leq 0$.*

(3) *Let* $\{R_t\}_{t\in T}$ *be any set of extremal rays of* $\overline{\mathrm{NE}}(X/Y;W)$. *Then*

$$\mathcal{N}_T := \{D \in \mathcal{L}(V;\pi^{-1}(W)) \mid (K_X + D)\cdot R_t \geq 0 \text{ for every } t \in T\}$$

is a rational polytope in V. *In particular,*

$$\mathcal{N}^\sharp_\pi(V;W) := \{\Delta \in \mathcal{L}(V;\pi^{-1}(W)) \mid K_X + \Delta \text{ is nef over } W\}$$

is a rational polytope.

We sometimes call $\mathcal{N}^\sharp_\pi(V;W)$ the *Shokurov polytope.*

Proof of Theorem 5.2.3. Throughout this proof, we can freely shrink Y around W suitably. We first note that $\mathcal{L}(V;\pi^{-1}(W))$ is a rational polytope in V (see 2.1.10).

(1) If Δ is a $\mathbb{Q}$-divisor, then we may assume that $m(K_X+\Delta)$ is Cartier for some positive integer m by shrinking Y around W suitably. Therefore, the statement is obvious even if Γ is not extremal. Hence, from now on, we assume that Δ is not a $\mathbb{Q}$-divisor. Then we can write $K_X + \Delta = \sum_j a_j(K_X + D_j)$ as in Lemma 2.3.5. This means that a_j is a positive real number for every j with $\sum_j a_j = 1$ and that $D_j \in \mathcal{L}(V;\pi^{-1}(W))$ is a $\mathbb{Q}$-divisor for every j. Thus we have $(K_X + \Delta)\cdot\Gamma = \sum_j a_j(K_X + D_j)\cdot\Gamma$. If $(K_X + \Delta)\cdot\Gamma < 1$, then

$$\begin{aligned} -2\dim X \leq (K_X + D_{j_0})\cdot\Gamma &< \frac{1}{a_{j_0}}\left\{-\sum_{j\neq j_0} a_j(K_X + D_j)\cdot\Gamma + 1\right\} \\ &\leq \frac{2\dim X + 1}{a_{j_0}} \end{aligned}$$

for $a_{j_0} \neq 0$. This is because $(K_X + D_j)\cdot\Gamma \geq -2\dim X$ holds for every j by Lemma 5.2.2. Thus there are only finitely many possibilities of the intersection numbers $(K_X + D_j)\cdot\Gamma$ for $a_j \neq 0$ when $(K_X + \Delta)\cdot\Gamma < 1$. Therefore, the existence of α is obvious.

(2) If we take δ sufficiently small, then, for every $D \in \mathcal{L}(V;\pi^{-1}(W))$ with $\|D - \Delta\| < \delta$, we can always find $D' \in \mathcal{L}(V;\pi^{-1}(W))$ such that

$$K_X + D = (1-s)(K_X + \Delta) + s(K_X + D')$$

with

$$0 \leq s \leq \frac{\alpha}{\alpha + 2\dim X}.$$

Since Γ is extremal, we have $(K_X+D')\cdot\Gamma \geq -2\dim X$ for every $D' \in \mathcal{L}(V;\pi^{-1}(W))$ by Lemma 5.2.2. We assume that $(K_X + \Delta)\cdot\Gamma > 0$. Then $(K_X + \Delta)\cdot\Gamma > \alpha$ by (1). Therefore,

$$\begin{aligned} (K_X + D)\cdot\Gamma &= (1-s)(K_X + \Delta)\cdot\Gamma + s(K_X + D')\cdot\Gamma \\ &> (1-s)\alpha + s(-2\dim X) \geq 0. \end{aligned}$$

This is a contradiction. Hence, we obtain $(K_X+\Delta)\cdot\Gamma\le 0$. We complete the proof of (2).

(3) For every $t\in T$, we may assume that there is some $D(t)\in\mathcal{L}(V;\pi^{-1}(W))$ such that $(K_X+D(t))\cdot R_t<0$. Let $B_1,\ldots,B_r$ be the vertices of $\mathcal{L}(V;\pi^{-1}(E))$. We note that $(K_X+D)\cdot R_t<0$ for some $D\in\mathcal{L}(V;\pi^{-1}(W))$ implies $(K_X+B_j)\cdot R_t<0$ for some j. Therefore, we may assume that T is contained in $\mathbb{N}$. This is because there are only countably many (K_X+B_j)-negative extremal rays for every j by the cone theorem (see Theorem 4.6.2). We note that $\mathcal{N}_T$ is a closed convex subset of $\mathcal{L}(V;\pi^{-1}(W))$ by definition. If T is a finite set, then the claim is obvious. Thus, we may assume that $T=\mathbb{N}$. By (2) and by the compactness of $\mathcal{N}_T$, we can take $\Delta_1,\ldots,\Delta_n\in\mathcal{N}_T$ and $\delta_1,\ldots,\delta_n>0$ such that $\mathcal{N}_T$ is covered by

$$\mathcal{B}_i=\left\{D\in\mathcal{L}(V;\pi^{-1}(W))\mid \|D-\Delta_i\|<\delta_i\right\}$$

and that if $D\in\mathcal{B}_i$ with $(K_X+D)\cdot R_t<0$ for some t, then $(K_X+\Delta_i)\cdot R_t=0$. If we put

$$T_i=\{t\in T\mid (K_X+D)\cdot R_t<0 \text{ for some } D\in\mathcal{B}_i\},$$

then $(K_X+\Delta_i)\cdot R_t=0$ for every $t\in T_i$ by the above construction. Since $\{\mathcal{B}_i\}_{i=1}^n$ gives an open covering of $\mathcal{N}_T$, we have $\mathcal{N}_T=\bigcap_{1\le i\le n}\mathcal{N}_{T_i}$ by the following claim.

Claim 1. $\mathcal{N}_T=\bigcap_{1\le i\le n}\mathcal{N}_{T_i}$.

Proof of Claim 1. We note that $\mathcal{N}_T\subset\bigcap_{1\le i\le n}\mathcal{N}_{T_i}$ is obvious. Suppose that $\mathcal{N}_T\subsetneq\bigcap_{1\le i\le n}\mathcal{N}_{T_i}$ holds. We take $D\in\bigcap_{1\le i\le n}\mathcal{N}_{T_i}\setminus\mathcal{N}_T$ which is very close to $\mathcal{N}_T$. Since $\mathcal{N}_T$ is covered by $\{\mathcal{B}_i\}_{i=1}^n$, there is some i_0 such that $D\in\mathcal{B}_{i_0}$. Since $D\notin\mathcal{N}_T$, there is some $t_0\in T$ such that $(K_X+D)\cdot R_{t_0}<0$. Thus, $t_0\in T_{i_0}$. This is a contradiction because $D\in\mathcal{N}_{T_{i_0}}$. Therefore, we obtain the desired equality $\mathcal{N}_T=\bigcap_{1\le i\le n}\mathcal{N}_{T_i}$. □

Therefore, it is sufficient to see that each $\mathcal{N}_{T_i}$ is a rational polytope in V. By replacing T with T_i, we may assume that there is some $D\in\mathcal{N}_T$ such that $(K_X+D)\cdot R_t=0$ for every $t\in T$.

Claim 2. *If* $\dim_{\mathbb{R}}\mathcal{L}(V;\pi^{-1}(W))=1$, *then* $\mathcal{N}_T$ *is a rational polytope in* V.

Proof of Claim 2. As we explained above, we can take some $D\in\mathcal{N}_T$ such that $(K_X+D)\cdot R_t=0$ for every $t\in T$. Since $\dim_{\mathbb{R}}\mathcal{L}(V;\pi^{-1}(W))=1$, we can write

$$K_X+D=b_1(K_X+D_1)+b_2(K_X+D_2)$$

such that $K_X+D_i\in\mathcal{L}(V;\pi^{-1}(W))$, D_i is a $\mathbb{Q}$-divisor, and $0\le b_i\le 1$ for $i=1,2$, and $b_1+b_2=1$. By $(K_X+D)\cdot R_t=0$, we see that b_1 and b_2 are rational numbers. This implies that D is a $\mathbb{Q}$-divisor. Therefore, $\mathcal{N}_T$ is a rational polytope in V. □

Hence we assume $\dim_{\mathbb{R}}\mathcal{L}(V;\pi^{-1}(W))>1$. Let $\mathcal{L}^1,\ldots,\mathcal{L}^p$ be the proper faces of $\mathcal{L}(V;\pi^{-1}(W))$. Then $\mathcal{N}_T^i=\mathcal{N}_T\cap\mathcal{L}^i$ is a rational polytope by induction on dimension. Moreover, for each $D'' \in \mathcal{N}_T$ which is not D, there is D' on some proper face of $\mathcal{L}(V; \pi^{-1}(W))$ such that D'' is on the line segment determined by D and D'. Note that $(K_X + D) \cdot R_t = 0$ for every $t \in T$. Therefore, if $D' \in \mathcal{L}^i$, then $D' \in \mathcal{N}^i_T$. Thus, $\mathcal{N}_T$ is the convex hull of D and all the $\mathcal{N}^i_T$. There is a finite subset $T' \subset T$ such that

$$\bigcup_i \mathcal{N}^i_T = \mathcal{N}_{T'} \cap (\bigcup_i \mathcal{L}^i).$$

Therefore, the convex hull of D and $\bigcup_i \mathcal{N}^i_T$ is just $\mathcal{N}_{T'}$. We complete the proof of (3). □

As an application of Theorem 5.2.3, we have:

Theorem 5.2.4. *Let $\pi\colon X \to Y$ be a projective morphism of complex analytic spaces such that X is a normal complex variety, and let W be a compact subset of Y such that the dimension of $N^1(X/Y; W)$ is finite. Let (X, Δ) be a log canonical pair and let H be an effective $\mathbb{R}$-Cartier $\mathbb{R}$-divisor on X such that $(X, \Delta + H)$ is log canonical and that $K_X + \Delta + H$ is nef over W. Then, either $K_X + \Delta$ is nef over W or there is a $(K_X + \Delta)$-negative extremal ray R of $\overline{\mathrm{NE}}(X/Y; W)$ such that $(K_X + \Delta + \lambda H) \cdot R = 0$, where*

$$\lambda := \inf \{t \geq 0 \mid K_X + \Delta + tH \textit{ is nef over } W\}.$$

Of course, $K_X + \Delta + \lambda H$ is nef over W.

The following picture helps the reader understand Theorem 5.2.4.

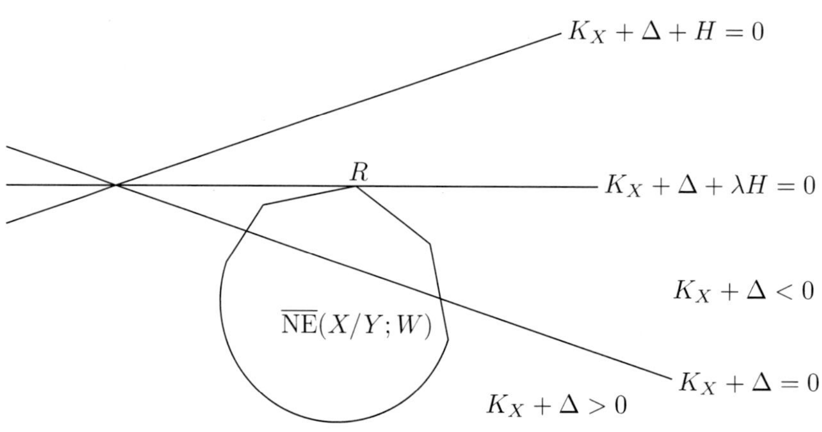

Proof of Theorem 5.2.4. We assume that $K_X + \Delta$ is not π-nef over W. Let $\{R_j\}$ be the set of the $(K_X + \Delta)$-negative extremal rays of $\overline{\mathrm{NE}}(X/Y; W)$. Let C_j be an extremal curve over W spanning R_j for every j. We put $\mu = \sup_j \{\mu_j\}$, where

$$\mu_j = \frac{-(K_X + \Delta) \cdot C_j}{H \cdot C_j}.$$

By definition, it is obvious that $\lambda = \mu$ and $0 < \mu \leq 1$ hold. Hence it is sufficient to prove that $\mu = \mu_{j_0}$ for some j_0. By Lemma 2.3.5, after shrinking Y around W suitably, we can find effective $\mathbb{Q}$-divisors $\Delta_1, \ldots, \Delta_k$ and positive real numbers $r_1, \ldots, r_k$ with $\sum_{i=1}^{k} r_i = 1$ such that $m(K_X + \Delta_i)$ is Cartier for every i, $\Delta = \sum_{i=1}^{k} r_i \Delta_i$ holds, and (X, Δ_i) is log canonical for every i. Therefore, by Lemma 5.2.2, we can write

$$-(K_X + \Delta) \cdot C_j = \sum_{i=1}^{l} \frac{r_i n_{ij}}{m} > 0,$$

where n_{ij} is an integer with $n_{ij} \leq 2m \dim X$ for every i and j since C_j is extremal over W. If $(K_X + \Delta + H) \cdot R_{j_0} = 0$ for some j_0, then there are nothing to prove since $\lambda = 1$ and $(K_X + \Delta + H) \cdot R = 0$ with $R = R_{j_0}$. Thus, we assume that $(K_X + \Delta + H) \cdot R_j > 0$ for every j. We put $F = \mathrm{Supp}(\Delta + H)$. Let $F = \sum_k F_k$ be the irreducible decomposition. We put $V = \bigoplus_k \mathbb{R}F_k$,

$$\mathcal{L}(V; \pi^{-1}(W)) := \left\{ D \in V \mid (X, D) \text{ is log canonical at } \pi^{-1}(W) \right\},$$

and

$$\mathcal{N} := \left\{ D \in \mathcal{L}(V; \pi^{-1}(W)) \mid (K_X + D) \cdot R_j \geq 0 \text{ for every } j \right\}.$$

Then $\mathcal{N}$ is a rational polytope in V by Theorem 5.2.3 (3) and $\Delta + H$ is in the relative interior of $\mathcal{N}$ by the above assumption. Therefore, after shrinking Y around W suitably again, we can write

$$K_X + \Delta + H = \sum_{p=1}^{q} r'_p (K_X + D_p),$$

where $r'_1, \ldots, r'_q$ are positive real numbers such that $\sum_p r'_p = 1$, (X, D_p) is log canonical for every p, $m'(K_X + D_p)$ is Cartier for some positive integer m' and every p, and $(K_X + D_p) \cdot C_j > 0$ for every p and j. So, we obtain

$$(K_X + \Delta + H) \cdot C_j = \sum_{p=1}^{q} \frac{r'_p n'_{pj}}{m'}$$

with $0 < n'_{pj} = m'(K_X + D_p) \cdot C_j \in \mathbb{Z}$. Note that m' and r'_p are independent of j for every p. We also note that

$$\begin{aligned} \frac{1}{\mu_j} = \frac{H \cdot C_j}{-(K_X + \Delta) \cdot C_j} &= \frac{(K_X + \Delta + H) \cdot C_j}{-(K_X + \Delta) \cdot C_j} + 1 \\ &= \frac{m \sum_{p=1}^{q} r'_p n'_{pj}}{m' \sum_{i=1}^{l} r_i n_{ij}} + 1. \end{aligned}$$

Since

$$\sum_{i=1}^{l} \frac{r_i n_{ij}}{m} > 0$$

for every j and $n_{ij} \leq 2m \dim X$ with $n_{ij} \in \mathbb{Z}$ for every i and j, the number of the set $\{n_{ij}\}_{i,j}$ is finite. Thus,

$$\inf_j \left\{ \frac{1}{\mu_j} \right\} = \frac{1}{\mu_{j_0}}$$

for some j_0. Therefore, we obtain $\mu = \mu_{j_0}$. We finish the proof. □

5.3 Basepoint-free theorem for $\mathbb{R}$-divisors

In this section, we will discuss the basepoint-free theorem for $\mathbb{R}$-divisors, although we do not need it in this book. The proof of Theorem 5.3.1 needs the cone theorem (see Theorem 4.6.2). Hence Theorem 5.3.1 looks much deeper than the basepoint-free theorem for Cartier divisors (see Theorem 4.2.1).

Theorem 5.3.1 (Basepoint-free theorem for $\mathbb{R}$-divisors). *Let $\pi\colon X \to Y$ be a projective morphism of complex analytic spaces such that X is a normal complex variety and let W be a compact subset of Y such that the dimension of $N^1(X/Y;W)$ is finite. Let Δ be an effective $\mathbb{R}$-divisor on X such that (X,Δ) is log canonical. Let D be an $\mathbb{R}$-Cartier $\mathbb{R}$-divisor defined on some open neighborhood of $\pi^{-1}(W)$ such that D is π-nef over W. Assume that $aD-(K_X+\Delta)$ is π-ample over W for some positive real number a. Then there exists an open neighborhood U of W such that D is semiample over U.*

Theorem 5.3.1 is an application of the cone theorem (see Theorem 4.6.2) and the basepoint-free theorem for Cartier divisors (see Theorem 4.2.1).

Proof of Theorem 5.3.1. Without loss of generality, by replacing D with aD, we may assume that $a=1$ holds. By replacing Y with a relatively compact open neighborhood of W, we may further assume that $\operatorname{Supp} D$ has only finitely many irreducible components and that D is a globally $\mathbb{R}$-Cartier $\mathbb{R}$-divisor on X. We consider

$$F = \left\{ z \in \overline{\mathrm{NE}}(X/Y;W) \mid D \cdot z = 0 \right\}.$$

Then F is a face of $\overline{\mathrm{NE}}(X/Y;W)$ and $(K_X+\Delta)\cdot z<0$ holds for $z \in F$. We take an ample $\mathbb{R}$-line bundle $\mathcal{A}$ on X such that $D-(K_X+\Delta+\mathcal{A})$ is still π-ample over W. Let R be a $(K_X+\Delta)$-negative extremal ray of $\overline{\mathrm{NE}}(X/Y;W)$ such that $R \subset F$. Then R is automatically a $(K_X+\Delta+\mathcal{A})$-negative extremal ray of $\overline{\mathrm{NE}}(X/Y;W)$ since $D\cdot R=0$ and $D-(K_X+\Delta+\mathcal{A})$ is π-ample over W. Therefore, F contains only finitely many $(K_X+\Delta)$-negative extremal rays $R_1,\ldots,R_k$ of $\overline{\mathrm{NE}}(X/Y;W)$. Thus, F is spanned by the extremal rays $R_1,\ldots,R_k$. Let $\operatorname{Supp} D=\sum_j D_j$ be the irreducible decomposition of $\operatorname{Supp} D$. Then we consider the finite-dimensional real vector space $V=\bigoplus_j \mathbb{R}D_j$. In this situation, we can easily check that

$$\mathcal{R} := \{B \in V \mid B \text{ is a globally } \mathbb{R}\text{-Cartier } \mathbb{R}\text{-divisor and } B\cdot z=0 \text{ for every } z \in F\}$$

is a rational affine subspace of V with $D \in \mathcal{R}$. As in Step 5 in the proof of Theorem 4.6.2, we put

$$\mathcal{C}_F := \overline{\mathrm{NE}}(X/Y; W)_{(K_X+\Delta+\mathcal{A})\geq 0} + \sum_{R_j \not\subset F} R_j.$$

and

$$\mathcal{R}^+ := \{B \in \mathcal{R} \mid B \text{ is positive on } \mathcal{C}_F \setminus \{0\}\}.$$

We note that $\overline{\mathrm{NE}}(X/Y; W)_{\mathrm{Nlc}(X,\Delta)} = \emptyset$ since (X, Δ) is log canonical. Then $\mathcal{R}^+$ is a non-empty open subset of $\mathcal{R}$ with $D \in \mathcal{R}^+$. Hence we can find positive real numbers $r_1, r_2, \ldots, r_m$ and globally $\mathbb{Q}$-Cartier $\mathbb{Q}$-divisors $B_1, B_2, \ldots, B_m \in \mathcal{R}^+$ such that $D = \sum_{i=1}^m r_i B_i$ and $B_i - (K_X + \Delta)$ is π-ample over W for every i. We note that B_i is automatically π-nef over W for every i since $B_i \in \mathcal{R}^+$. By the basepoint-free theorem for Cartier divisors (see Theorem 4.2.1), there exists a relatively compact open neighborhood U of W such that B_i is π-semiample over U for every i. Therefore, $D = \sum_{i=1}^m r_i B_i$ is π-semiample over U. This is what we wanted. □

Theorem 5.3.1 will play an important role in the study of minimal models of complex analytic spaces.

5.4 Strong $\mathbb{Q}$-factoriality

In this section, we will discuss strong $\mathbb{Q}$-factoriality. It is indispensable for the minimal model program of normal complex varieties since a divisorial sheaf on a normal complex variety does not always come from a Weil divisor. Let us start with the definition of *divisorial sheaves*.

Definition 5.4.1 (Divisorial sheaves)**.** Let X be a normal complex variety. A coherent sheaf $\mathcal{L}$ on X is said to be *divisorial* if it is reflexive of rank one.

The following remark is very important and is well known.

Remark 5.4.2. Let X and $\mathcal{L}$ be algebraic in Definition 5.4.1. Then there always exists a Weil divisor D on X such that $\mathcal{L} = \mathcal{O}_X(D)$.

Lemma 5.4.3. *Let $\mathcal{L}$ be a divisorial sheaf on a smooth complex variety. Then $\mathcal{L}$ is a line bundle, that is, $\mathcal{L}$ is invertible.*

Proof. This fact is very well known. For the proof, see, for example, [OSS, Chapter 2, Lemma 1.1.15]. □

Following [DH], we define *strongly $\mathbb{Q}$-factorial* normal complex varieties.

Definition 5.4.4 (Strong $\mathbb{Q}$-factoriality)**.** Let $\mathcal{L}$ be a divisorial sheaf on a normal complex variety X. Let x be a point of X. If there exists a positive integer m_x such that

$$\mathcal{L}^{[m_x]} := \left(\mathcal{L}^{\otimes m_x}\right)^{**}$$

is a line bundle on some open neighborhood of $x \in X$, then $\mathcal{L}$ is said to be $\mathbb{Q}$*-Cartier at* x. Let K be a subset of X. If $\mathcal{L}$ is $\mathbb{Q}$-Cartier at every $x \in K$, then $\mathcal{L}$ is said to be $\mathbb{Q}$*-Cartier at* K. When $K = X$, we simply say that $\mathcal{L}$ is $\mathbb{Q}$*-Cartier*.

We say that X is *strongly* $\mathbb{Q}$*-factorial at* K if every divisorial sheaf $\mathcal{L}$ defined over some open neighborhood $\mathcal{U}_{\mathcal{L}}$ of K is $\mathbb{Q}$-Cartier at K. As above, when $K = X$, we simply say that X is *strongly* $\mathbb{Q}$*-factorial*.

Let $\pi \colon X \to Y$ be a morphism of complex analytic spaces and let W be a subset of Y. If X is strongly $\mathbb{Q}$-factorial at $\pi^{-1}(W)$, then we say that X is *strongly* $\mathbb{Q}$*-factorial over* W.

Remark 5.4.5. Let X be a normal complex variety and let K and L be subsets of X with $L \subsetneq K$. We note that X is not necessarily strongly $\mathbb{Q}$-factorial at L even when X is strongly $\mathbb{Q}$-factorial at K.

We are mainly interested in the case where K is a compact subset of X in Definition 5.4.4.

Lemma 5.4.6. *Let* $\mathcal{L}$ *be a divisorial sheaf on a normal complex variety* X *and let* K *be a compact subset of* X*. If* $\mathcal{L}$ *is* $\mathbb{Q}$*-Cartier at* K*, then there exists a positive integer* m *such that* $\mathcal{L}^{[m]}$ *is a line bundle on some open neighborhood of* K*.*

Proof. By definition, we can take a finite open cover $\{\mathcal{U}_i\}_{i\in I}$ of K such that $\mathcal{L}^{[m_i]}$ is a line bundle on $\mathcal{U}_i$, where m_i is a positive integer, for every $i \in I$ since K is compact. We put $m := \operatorname{lcm}_{i\in I}\{m_i\}$. Then it is easy to see that $\mathcal{L}^{[m]}$ is a line bundle on $\mathcal{U} := \bigcup_{i\in I} \mathcal{U}_i$. This is what we wanted. □

Lemma 5.4.7. *A smooth complex variety is strongly* $\mathbb{Q}$*-factorial.*

Proof. This is obvious by Lemma 5.4.3. □

The following lemma is more or less well known. We state it here for the sake of completeness.

Lemma 5.4.8. *Let* $f \colon X \to Y$ *be a small projective bimeromorphic morphism of normal complex varieties. Assume that* Y *is strongly* $\mathbb{Q}$*-factorial at* $y \in Y$*. Then* f *is an isomorphism over some open neighborhood of* $y \in Y$*.*

Proof. We take an f-ample line bundle $\mathcal{A}$ on X. We put $\mathcal{L} := (f_*\mathcal{A})^{**}$. Then $\mathcal{L}$ is a divisorial sheaf on Y. By assumption, there exists a positive integer m such that $\mathcal{L}^{[m]}$ is a line bundle on some open neighborhood U of $y \in Y$. Since f is small, we can easily see that $\mathcal{A}^{\otimes m} = f^*\mathcal{L}^{[m]}$ holds over U. This implies that f is an isomorphism over U because $\mathcal{A}$ is f-ample. □

When we run the minimal model program, we need the following two lemmas. The proof is essentially the same as for algebraic varieties. We explain it in detail for the sake of completeness.

Lemma 5.4.9 (Divisorial contractions). *Let $\pi\colon X \to Y$ be a projective morphism of complex analytic spaces and let W be a compact subset of Y. Let (X, Δ) be a log canonical pair. Let $\varphi_R\colon X \to Z$ be the divisorial contraction associated to a $(K_X + \Delta)$-negative extremal ray R of $\overline{\mathrm{NE}}(X/Y; W)$. Assume that X is strongly $\mathbb{Q}$-factorial over W. Then Z is also strongly $\mathbb{Q}$-factorial over W.*

Proof. Throughout this proof, we will freely shrink Y around W without mentioning it explicitly. Let $\mathcal{L}$ be a divisorial sheaf on Z. We put $\mathcal{M} := (\varphi_R^*\mathcal{L})^{**}$. Then $\mathcal{M}$ is a divisorial sheaf on X. Let E be the φ_R-exceptional divisor on X. We can take a positive integer m and an integer r such that $\mathcal{M}^{[m]}$ and $\mathcal{O}_X(rE)$ are line bundles on X and $\left(\mathcal{M}^{[m]} \otimes \mathcal{O}_X(rE)\right) \cdot C = 0$ for every $[C] \in R$. By the cone and contraction theorem for log canonical pairs (see Theorems 4.6.1 and 4.6.2), there exists a line bundle $\mathcal{N}$ on Z such that $\mathcal{M}^{[m]} \otimes \mathcal{O}_X(rE) \simeq \varphi_R^*\mathcal{N}$. This implies that $\mathcal{L}^{[m]} \simeq \mathcal{N}$. Hence $\mathcal{L}$ is $\mathbb{Q}$-Cartier. Thus Z is strongly $\mathbb{Q}$-factorial over W. We finish the proof. □

Lemma 5.4.10 (Flips). *Let $\pi\colon X \to Y$ be a projective morphism of complex analytic spaces and let W be a compact subset of Y. Let (X, Δ) be a log canonical pair. Assume that X is strongly $\mathbb{Q}$-factorial over W. Let $\varphi_R\colon X \to Z$ be the flipping contraction associated to a $(K_X + \Delta)$-negative extremal ray R of $\overline{\mathrm{NE}}(X/Y; W)$ and let $\varphi_R^+\colon X^+ \to Z$ be the flip of φ_R.*

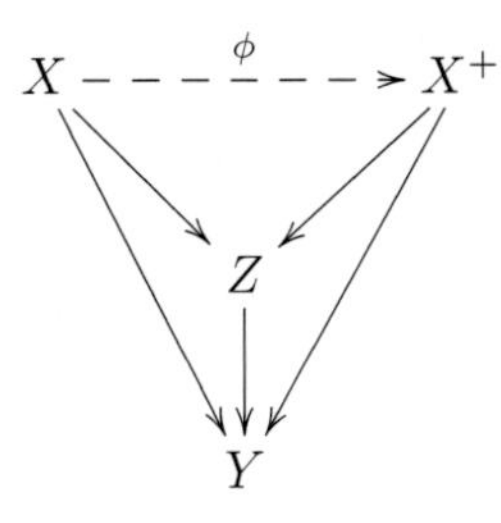

Then X^+ is also strongly $\mathbb{Q}$-factorial over W.

Proof. Throughout this proof, we will freely shrink Y around W without mentioning it explicitly. By perturbing the coefficients of Δ, we may assume that $K_X + \Delta$ is $\mathbb{Q}$-Cartier. Let $\mathcal{L}$ be a divisorial sheaf on X^+. Since $\phi\colon X \dashrightarrow X^+$ is an isomorphism in codimension one, we can define a divisorial sheaf $\mathcal{M}$ on X which is the strict transform of $\mathcal{L}$. Then there exist a positive integer m and an integer r such that $\mathcal{M}^{[m]}$ and $\mathcal{O}_X(r(K_X + \Delta))$ are line bundles and $\left(\mathcal{M}^{[m]} \otimes \mathcal{O}_X(r(K_X + \Delta))\right) \cdot C = 0$ holds for every $[C] \in R$. By the cone and contraction theorem for log canonical pairs (see Theorems 4.6.1 and 4.6.2), this implies that there exists a line bundle $\mathcal{N}$ on Z such that $\mathcal{M}^{[m]} \otimes \mathcal{O}_X(r(K_X + \Delta)) \simeq \varphi_R^*\mathcal{N}$. Hence we have $\mathcal{L}^{[m]} \otimes \mathcal{O}_X(r(K_{X^+} + \Delta^+)) \simeq (\varphi_R^+)^*\mathcal{N}$, where Δ^+ is the strict transform of Δ on X^+. Therefore, $\mathcal{L}$ is $\mathbb{Q}$-Cartier over W. Thus, X^+ is strongly $\mathbb{Q}$-factorial over W. We finish the proof. □

By Lemmas 5.4.9 and 5.4.10, we see that the strong $\mathbb{Q}$-factoriality is preserved under the minimal model program starting from a strongly $\mathbb{Q}$-factorial log canonical pairs.

5.5 Proof of Main theorem: Theorem 1.1.6

In this section, we will prove Theorem 1.1.6, which is the main theorem of this book.

Proof of Theorem 1.1.6. By Theorem 4.6.2 (1) and (2), we obtain the following equality

$$\overline{\mathrm{NE}}(X/Y;W) = \overline{\mathrm{NE}}(X/Y;W)_{(K_X+\Delta)\geq 0} + \overline{\mathrm{NE}}(X/Y;W)_{\mathrm{Nlc}(X,\Delta)} + \sum R_j$$

satisfying (1), (2), and (3) in Theorem 1.1.6. By Theorem 4.6.2 (3), F in Theorem 1.1.6 (4) is rational and hence contractible at $\mathrm{Nlc}(X,\Delta)$. Thus, by the contraction theorem (see Theorem 4.6.1), we obtain the desired contraction morphism $\varphi_F\colon X \to Z$ over Y after shrinking Y around W suitably. By Theorem 5.1.3, we see that (5) holds. We note that (6) is nothing but Theorem 5.2.4. Finally, we will prove (7). We may assume that $K_X+\Delta$ is not π-nef over W. Then we can take a small positive real number ε such that $K_X+\Delta+\varepsilon\mathcal{H}$ is not π-nef over W. By the cone theorem (see Theorem 4.6.2 (2)), there exist only finitely many $(K_X+\Delta+\varepsilon\mathcal{H})$-negative extremal rays $R_1,\dots,R_k$ of $\overline{\mathrm{NE}}(X/Y;W)$. We put

$$\mu_i := \frac{-(K_X+\Delta)\cdot R_i}{\mathcal{H}\cdot R_i}$$

for every i. Then it is obvious that $\lambda = \max_{1\leq i\leq k}\mu_i$. If $\lambda = \mu_{i_0}$ holds for $1\leq i_0\leq k$, then $(K_X+\Delta+\lambda\mathcal{H})\cdot R_{i_0} = 0$. By construction, $K_X+\Delta+\lambda\mathcal{H}$ is π-nef over W. We finish the proof of Theorem 1.1.6. $\square$

Chapter 6
Advanced topics

In this final chapter, we will discuss some advanced topics. In Section 6.1, we will quickly review the theory of quasi-log complex analytic spaces without proof. The reader can find all the details in [Fu16]. In Section 6.2, we will explain the basepoint-free theorem of Reid–Fukuda type for log canonical pairs in the complex analytic setting (see Theorem 6.2.1). The proof of Theorem 6.2.1 needs the framework of quasi-log complex analytic spaces and is much harder than the one for the usual basepoint-free theorem (see Theorem 4.2.1). In Section 6.3, we will review the effective freeness and very ampleness for log canonical pairs in the complex analytic setting. In Section 6.4, we will explain some results on the minimal model program for projective morphisms between complex analytic spaces established in [Fu13].

6.1 On quasi-log structures

In this section, we will briefly explain the theory of quasi-log complex analytic spaces without proof. In [Fu16], we will discuss quasi-log structures for complex analytic spaces in detail. For the details of the theory of *quasi-log schemes*, see [Fu6, Chapter 6]. Let us start with the definition of *quasi-log complex analytic spaces*.

Definition 6.1.1 (Quasi-log complex analytic spaces). A *quasi-log complex analytic space*

$$\left(X, \omega, f\colon (Z, B_Z) \to X\right)$$

is a complex analytic space X endowed with an $\mathbb{R}$-line bundle (or a globally $\mathbb{R}$-Cartier divisor) ω on X, a closed analytic subspace $X_{-\infty} \subsetneq X$, and a finite collection $\{C\}$ of reduced and irreducible closed analytic subspaces of X such that there exists a projective morphism $f\colon (Z, B_Z) \to X$ from an analytic globally embedded simple normal crossing pair (Z, B_Z) satisfying the following properties:

(i) $f^*\omega \sim_{\mathbb{R}} K_Z + B_Z$.

(ii) The natural map $\mathcal{O}_X \to f_*\mathcal{O}_Z(\lceil -(B_Z^{<1})\rceil)$ induces an isomorphism

$$\mathcal{I}_{X_{-\infty}} \xrightarrow{\simeq} f_*\mathcal{O}_Z(\lceil -(B_Z^{<1})\rceil - \lfloor B_Z^{>1}\rfloor),$$

where $\mathcal{I}_{X_{-\infty}}$ is the defining ideal sheaf of $X_{-\infty}$ on X.

(iii) The collection of closed analytic subvarieties $\{C\}$ coincides with the f-images of (Z, B_Z)-strata that are not included in $X_{-\infty}$.

We often simply write $[X, \omega]$ to denote the above data

$$\left(X, \omega, f\colon (Z, B_Z) \to X\right)$$

if there is no risk of confusion. The closed analytic subvarieties C are called the *qlc strata* of $[X, \omega]$. If a qlc stratum C is not an irreducible component of X, then it is called a *qlc center* of $[X, \omega]$. The closed analytic subspace $X_{-\infty}$ is called the *non-qlc locus* of $[X, \omega]$. We note that we sometimes use $\mathrm{Nqlc}(X, \omega)$ or $\mathrm{Nqlc}\,(X, \omega, f\colon (Z, B_Z) \to X)$ to denote $X_{-\infty}$.

Furthermore, if $X_{-\infty} = \emptyset$ in the above definition, then

$$(X, \omega, f\colon (Z, B_Z) \to X)$$

is called a *quasi-log canonical pair*.

For the details of Definition 6.1.1, see [Fu16, Definition 4.1]. The following example is the most important one.

Example 6.1.2 (Normal pairs). Let $\pi\colon X \to Y$ be a projective morphism of complex analytic spaces such that X is a normal complex variety, and let Δ be an effective $\mathbb{R}$-divisor on X such that $K_X + \Delta$ is $\mathbb{R}$-Cartier. We usually call (X, Δ) a *normal pair*. We replace Y with any relatively compact open subset of Y. Then we can construct a projective bimeromorphic morphism $f\colon Z \to X$ with

$$K_Z + B_Z := f^*(K_X + \Delta)$$

such that Z is smooth and $\mathrm{Supp}\, B_Z$ is a simple normal crossing divisor on Y. Then

$$f^*(K_X + \Delta) \sim_{\mathbb{R}} K_Z + B_Z$$

obviously holds and

$$\mathcal{J}_{\mathrm{NLC}}(X, \Delta) = f_*\mathcal{O}_Z(\lceil -(B_Z^{<1}) \rceil - \lfloor B_Z^{>1} \rfloor)$$

is a well-defined coherent ideal sheaf on X which defines the non-lc locus $\mathrm{Nlc}(X, \Delta)$ of (X, Δ) (see Definition 2.3.1). By definition, C is a log canonical center of (X, Δ) if and only if C is not contained in $\mathrm{Nlc}(X, \Delta)$ and is the f-image of some log canonical center of $(Z, B_Z^{=1})$. Hence,

$$(X, K_X + \Delta, f\colon (Z, B_Z) \to X)$$

with $X_{-\infty} := \mathrm{Nlc}(X, \Delta)$ satisfies the conditions in Definition 6.1.1, that is,

$$(X, K_X + \Delta, f\colon (Z, B_Z) \to X)$$

is a quasi-log complex analytic space. By construction, $X_{-\infty} = \emptyset$ if and only if (X, Δ) is log canonical.

By Example 6.1.2, we can use the theory of quasi-log complex analytic spaces to study normal pairs. A key result of the theory of quasi-log complex analytic spaces is *adjunction*.

Theorem 6.1.3 (Adjunction formula for quasi-log complex analytic spaces). *Let*

$$\left(X, \omega, f\colon (Z, B_Z) \to X\right)$$

be a quasi-log complex analytic space and let X' be the union of $X_{-\infty}$ with a union of some qlc strata of $[X, \omega]$. Then, after replacing X with any relatively compact open subset of X, we can construct a projective morphism $f'\colon (Z', B_{Z'}) \to X'$ from an analytic globally embedded simple normal crossing pair $(Z', B_{Z'})$ such that

$$\left(X', \omega', f'\colon (Z', B_{Z'}) \to X'\right)$$

is a quasi-log complex analytic space with $\omega' = \omega|_{X'}$ and $X'_{-\infty} = X_{-\infty}$. Moreover, the qlc strata of $[X', \omega']$ are exactly the qlc strata of $[X, \omega]$ that are included in X'.

By Theorem 6.1.3, we can use inductive arguments for the study of quasi-log complex analytic spaces. Note that Theorem 6.1.3 follows from Theorem 3.1.5 (1). By using Theorem 3.2.3, we can obtain the following vanishing theorems.

Theorem 6.1.4 (Vanishing theorem I). *Let*

$$\left(X, \omega, f\colon (Z, B_Z) \to X\right)$$

be a quasi-log complex analytic space and let $\pi\colon X \to Y$ be a projective morphism between complex analytic spaces. Let $\mathcal{L}$ be a line bundle on X such that $\mathcal{L} - \omega$ is nef and log big over Y with respect to $[X, \omega]$, that is, $\mathcal{L} - \omega$ is nef over Y and $(\mathcal{L} - \omega)|_C$ is big over $\pi(C)$ for every qlc stratum C of $[X, \omega]$. Then

$$R^i\pi_*(\mathcal{I}_{X_{-\infty}} \otimes \mathcal{L}) = 0$$

holds for every $i > 0$. In particular, if $X_{-\infty} = \emptyset$, then

$$R^i\pi_*\mathcal{L} = 0$$

holds for every $i > 0$.

Theorem 6.1.5 (Vanishing theorem II). *Let*

$$\left(X, \omega, f\colon (Z, B_Z) \to X\right)$$

be a quasi-log complex analytic space and let X' be the union of $X_{-\infty}$ with a union of some qlc strata of $[X, \omega]$. Let $\pi\colon X \to Y$ be a projective morphism between complex analytic spaces and let $\mathcal{L}$ be a line bundle on X such that $\mathcal{L} - \omega$ is nef over Y and $(\mathcal{L} - \omega)|_C$ is big over $\pi(C)$ for every qlc stratum C of $[X, \omega]$ which is not contained in X'. Then

$$R^i\pi_*(\mathcal{I}_{X'} \otimes \mathcal{L}) = 0$$

holds for every $i > 0$, where $\mathcal{I}_{X'}$ is the defining ideal sheaf of X' on X. In particular, if $\mathcal{L} - \omega$ is ample over Y, then

$$R^i\pi_*(\mathcal{I}_{X'} \otimes \mathcal{L}) = 0$$

holds for every $i > 0$.

We can see Theorem 6.1.5 as a generalization of the Kawamata–Viehweg–Nadel vanishing theorem. For the details of the above theorems, see [Fu16, Section 4]. We note that Theorems 6.1.4 and 6.1.5 generalize the results in Section 3.3 by Example 6.1.2.

6.2 Basepoint-free theorem of Reid–Fukuda type

The following theorem is a special case of [Fu16, Theorem 7.1], where we formulate the basepoint-free theorem of Reid–Fukuda type for quasi-log complex analytic spaces.

Theorem 6.2.1 (Basepoint-free theorem of Reid–Fukuda type). *Let $\pi\colon X \to Y$ be a projective morphism of complex analytic space such that X is a normal complex variety, and let Δ be an effective $\mathbb{R}$-divisor on X such that $K_X + \Delta$ is $\mathbb{R}$-Cartier. Let $\mathcal{L}$ be a π-nef line bundle on X such that $a\mathcal{L} - (K_X + \Delta)$ is nef and log big over Y with respect to (X, Δ) for some positive real number a. This means that $a\mathcal{L} - (K_X + \Delta)$ is nef and big over Y and that $(a\mathcal{L} - (K_X + \Delta))|_C$ is big over $\pi(C)$ for every log canonical center C of (X, Δ). We assume that $\mathcal{L}^{\otimes m}|_{\mathrm{Nlc}(X,\Delta)}$ is π-generated for every $m \gg 0$. Then, after replacing Y with any relatively compact open subset of Y, $\mathcal{L}^{\otimes m}$ is π-generated for every $m \gg 0$.*

Theorem 6.2.1 obviously generalizes the basepoint-free theorem: Theorem 4.2.1. To the best knowledge of the author, there is no proof of Theorem 6.2.1 without using the theory of quasi-log complex analytic spaces. Hence Theorem 6.2.1 is much harder to prove than Theorem 4.2.1. For the details of Theorem 6.2.1, see [Fu16, Section 7].

6.3 Effective freeness and very ampleness

In this section, we will briefly explain the effective freeness and the effective very ampleness for log canonical pairs in the complex analytic setting without proof. The following theorem is a very special case of [Fu16, Theorem 8.1] by Example 6.1.2.

Theorem 6.3.1 (Effective freeness). *Let $\pi\colon X \to Y$ be a projective morphism between complex analytic spaces and let (X, Δ) be a log canonical pair. Let $\mathcal{L}$ be a π-nef line bundle on X such that $a\mathcal{L} - (K_X + \Delta)$ is π-ample over Y for some non-negative integer a. Then there exists a positive integer $m = m(\dim X, a)$, which only depends on $\dim X$ and a, such that $\mathcal{L}^{\otimes m}$ is π-generated. Moreover, there exists a positive integer $m_0 = m_0(\dim X, a)$ depending only on $\dim X$ and a such that $\mathcal{L}^{\otimes l}$ is π-generated for every $l \geq m_0$.*

If $K_X + \Delta$ is $\mathbb{Q}$-Cartier in Theorem 6.3.1, then we have:

Theorem 6.3.2. *In Theorem 6.3.1, if $K_X + \Delta$ is $\mathbb{Q}$-Cartier, then we may replace the assumption that $a\mathcal{L} - (K_X + \Delta)$ is π-ample over Y with a weaker one that $a\mathcal{L} - (K_X + \Delta)$ is nef and log big over Y with respect to (X, Δ), that is, $a\mathcal{L} - (K_X + \Delta)$ is* *nef and big over Y and $(a\mathcal{L} - (K_X + \Delta))|_C$ is big over $\pi(C)$ for every log canonical center C of (X, Δ).*

For the details of Theorem 6.3.2, see [Fu16, Theorem 8.2]. If $\mathcal{L}$ is π-ample in Theorem 6.3.1, then we have the following effective very ampleness.

Theorem 6.3.3 (Effective very ampleness). *Let $\pi: X \to Y$ be a projective morphism between complex analytic spaces and let (X, Δ) be a log canonical pair. Let $\mathcal{L}$ be a π-ample line bundle on X such that $a\mathcal{L} - (K_X + \Delta)$ is π-nef over Y for some non-negative integer a. Then there exists a positive integer $m' = m'(\dim X, a)$ depending only on $\dim X$ and a such that $\mathcal{L}^{\otimes m'}$ is π-very ample. Moreover, there exists a positive integer $m'_0 = m'_0(\dim X, a)$ depending only on $\dim X$ and a such that $\mathcal{L}^{\otimes l'}$ is π-very ample for every $l' \geq m'_0$.*

As above, Theorem 6.3.3 is a special case of [Fu16, Theorem 8.3] by Example 6.1.2. For the details see [Fu16, Section 8], where everything is formulated for quasi-log complex analytic spaces.

6.4 Minimal model program

In this section, we will shortly discuss some results established in [Fu13]. Roughly speaking, in [Fu13], we showed that [BCHM] and [HM] can work for projective morphisms between complex analytic spaces. Hence we have established the finite generation for kawamata log terminal pairs in the complex analytic setting.

Theorem 6.4.1 (Finite generation). *Let $\pi: X \to Y$ be a projective morphism of complex analytic spaces and let Δ be an effective $\mathbb{Q}$-divisor on X such that (X, Δ) is kawamata log terminal. Then*

$$R(X/Y, K_X + \Delta) := \bigoplus_{m \in \mathbb{N}} \pi_* \mathcal{O}_X(\lfloor m(K_X + \Delta) \rfloor)$$

is a locally finitely generated graded $\mathcal{O}_Y$-algebra.

As a special case of Theorem 6.4.1, we have the following finite generation of relative canonical rings.

Corollary 6.4.2. *Let $\pi: X \to Y$ be a projective morphism of complex analytic spaces such that X is smooth and irreducible. Then the relative canonical ring*

$$R(X/Y) := \bigoplus_{m \in \mathbb{N}} \pi_* \omega_X^{\otimes m}$$

is a locally finitely generated graded $\mathcal{O}_Y$-algebra.

In the complex analytic setting, we established the minimal model program with scaling as follows.

Theorem 6.4.3 (Minimal model program with scaling for projective morphisms of complex analytic spaces, see [Fu13, Theorem 1.7]). *Let $\pi: X \to Y$ be a projective surjective morphism of complex analytic spaces and let W be a compact subset of*

Y. Assume that Y is Stein and that W is a Stein compact subset of Y such that $\Gamma(W, \mathcal{O}_Y)$ is noetherian. Let (X, Δ) be a kawamata log terminal pair such that Δ is π-big and that X is $\mathbb{Q}$-factorial over W. If C is an effective $\mathbb{R}$-divisor on X such that $K_X + \Delta + C$ is kawamata log terminal and it is nef over W. Then we can run the $(K_X + \Delta)$-minimal model program with scaling of C over Y around W. More precisely, we have a finite sequence of flips and divisorial contractions over Y starting from (X, Δ):

$$(X, \Delta) =: (X_0, \Delta_0) \overset{\varphi_0}{\dashrightarrow} (X_1, \Delta_1) \overset{\varphi_1}{\dashrightarrow} \cdots \overset{\varphi_{m-1}}{\dashrightarrow} (X_m, \Delta_m),$$

where $\Delta_{i+1} := (\varphi_i)_*\Delta_i$ for every $i \geq 0$, such that (X_m, Δ_m) is a log terminal model (resp. has a Mori fiber space structure) over some open neighborhood of W when $K_X + \Delta$ is π-pseudo-effective (resp. not π-pseudo-effective). We note that each step φ_i exists only after shrinking Y around W suitably. Hence we have to replace Y with a small Stein open neighborhood of W repeatedly in the above process.

For the details of Theorem 6.4.3, see [Fu13]. After [Fu13], we formulated and proved necessary vanishing theorems in [Fu15] (see also Chapter 3, [FF], [Fu18], and so on). Then, in this book, we established the cone and contraction theorem for log canonical pairs in the complex analytic setting. This is a starting point of the minimal model program for log canonical complex analytic spaces. Recently, the minimal model program for log canonical pairs in the complex analytic setting was investigated in [EH1], [EH2], [Fu19], and [EH3]. Roughly speaking, we proved that almost all results in [Bir], [HasH], and [Has2] hold true even in the complex analytic setting. We strongly recommend the interested reader to see [EH1], [EH2], [Fu19], [EH3], and so on.

Chapter 7
Erratum to [Fu6]

Here, we will correct some small mistakes in [Fu6] for the reader's convenience. Let us correct some mathematical mistakes.

- The proof of [Fu6, Lemma 2.4.5] has a small trouble. It is not clear whether the pull-back of E on a resolution of X is still in the movable cone or not. Hence it may be better to add the assumption that X is $\mathbb{Q}$-factorial in [Fu6, Lemma 2.4.5]. Then, the proof of [Fu6, Lemma 2.4.5] works without replacing X with a resolution. Note that [Fu6, Lemma 2.4.5] is used in the proof of [Fu6, Theorem 4.4.21] and that Z in the proof of [Fu6, Theorem 4.4.21] is $\mathbb{Q}$-factorial. Therefore, it is harmless to add the assumption that X is $\mathbb{Q}$-factorial to [Fu6, Lemma 2.4.5].
- Lemma 2.4.9 in [Fu6] is not correct. There exists a counterexample (see [Le, Theorem 1]). For the details, see [Fu8]. This mistake affects Remark 4.11.6 in [Fu6]. Fortunately, it was rescued by [Fu8, Theorem 3.2]. Unfortunately, however, the proof of the inequality in [Fu6, Theorem 4.12.1] has not been completed yet. We have only established the following two weaker inequalities

  $$\kappa_\sigma(X, K_X + D_X) \geq \kappa_\sigma(F, K_F + D_X|_F) + \kappa(Y, K_Y + D_Y)$$

  and

  $$\kappa_\sigma(X, K_X + D_X) \geq \kappa(F, K_F + D_X|_F) + \kappa_\sigma(Y, K_Y + D_Y).$$

  For the details, see [Fu8, Theorem 2.1]. Hence, we have to correct [Fu6, Corollary 4.12.2] and the proof of [Fu6, Theorem 4.12.8] suitably. On [Na2, Chapter V, 4.1. Theorem (1)] (see [Fu6, Remark 4.11.6]), we strongly recommend the reader to see [Fu9, Theorem 1.12].

We correct some typos in [Fu6].

- On page 172, line 2,

  $$\sum_{i=1}^{n} d_i D_i$$

  should be replaced by

  $$\sum_{i=1}^{r} d_i D_i.$$

- On page 227, line 10, $X' = f(Y')$ should be replaced by $f(Y')$. In general, $X' = f(Y')$ does not always hold.

- On page 237, line 22,

$$\pi_*\mathcal{O}_X(mL)) \to \pi_*\mathcal{O}_{X'}(mL)$$

should be replaced by

$$\pi_*\mathcal{O}_X(mL) \to \pi_*\mathcal{O}_{X'}(mL).$$

- On page 245, line 17, $z, z' \in F$ should be replaced by $z, z' \in \overline{NE}(X/S)$.

References

[BăS] C. Bănică, O. Stănăşilă, *Algebraic methods in the global theory of complex spaces*, Translated from the Romanian. Editura Academiei, Bucharest; John Wiley & Sons, London–New York–Sydney, 1976.

[BieM1] E. Bierstone, P. D. Milman, Semianalytic and subanalytic sets, Inst. Hautes Études Sci. Publ. Math. No. **67** (1988), 5–42.

[BieM2] E. Bierstone, P. D. Milman, Canonical desingularization in characteristic zero by blowing up the maximum strata of a local invariant, Invent. Math. **128** (1997), no. 2, 207–302.

[Bir] C. Birkar, Existence of log canonical flips and a special LMMP, Publ. Math. Inst. Hautes Études Sci. **115** (2012), 325–368.

[BCHM] C. Birkar, P. Cascini, C. D. Hacon, J. McKernan, Existence of minimal models for varieties of log general type, J. Amer. Math. Soc. **23** (2010), no. 2, 405–468.

[CL] P. Cascini, V. Lazić, New outlook on the minimal model program, I, Duke Math. J. **161** (2012), no. 12, 2415–2467.

[DH] O. Das, C. D. Hacon, The log minimal model program for Kähler 3-folds, preprint (2020). arXiv:2009.05924 [math.AG]

[DHP] O. Das, C. D. Hacon, M. Păun, On the 4-dimensional minimal model program for Kähler varieties, Adv. Math. **443** (2024), Paper No. 109615, 68 pp.

[EH1] M. Enokizono, K. Hashizume, Semistable reduction for complex analytic spaces, preprint (2024). arXiv:2404.05124 [math.AG]

[EH2] M. Enokizono, K. Hashizume, Minimal model program for log canonical pairs on complex analytic spaces, preprint (2024). arXiv:2404.05126 [math.AG]

[EH3] M. Enokizono, K. Hashizume, On termination of minimal model program for log canonical pairs on complex analytic spaces, preprint (2024).

[Fi] G. Fischer, *Complex analytic geometry*, Lecture Notes in Mathematics, Vol. **538**. Springer-Verlag, Berlin–New York, 1976.

[Fu1] O. Fujino, Abundance theorem for semi log canonical threefolds, Duke Math. J. 102 (2000), no. 3, 513–532.

[Fu2] O. Fujino, What is log terminal?, *Flips for 3-folds and 4-folds*, 49–62, Oxford Lecture Ser. Math. Appl., **35**, Oxford Univ. Press, Oxford, 2007.

[Fu3] O. Fujino, On Kawamata's theorem, *Classification of algebraic varieties*, 305–315, EMS Ser. Congr. Rep., Eur. Math. Soc., Zürich, 2011.

[Fu4] O. Fujino, Fundamental theorems for the log minimal model program, Publ. Res. Inst. Math. Sci. **47** (2011), no. 3, 727–789.

[Fu5] O. Fujino, A transcendental approach to Kollár's injectivity theorem II, J. Reine Angew. Math. **681** (2013), 149–174.

[Fu6] O. Fujino. *Foundations of the minimal model program*, MSJ Memoirs, **35**. Mathematical Society of Japan, Tokyo, 2017.

[Fu7] O. Fujino, Pull-back of quasi-log structures, Publ. Res. Inst. Math. Sci. **53** (2017), no. 2, 241–259.

[Fu8] O. Fujino, Corrigendum to "On subadditivity of the logarithmic Kodaira dimension", J. Math. Soc. Japan **72** (2020), no. 4, 1181–1187.

[Fu9] O. Fujino, On mixed ω-sheaves, Asian J. Math. **28** (2024), no. 2, 197–242.

[Fu10] O. Fujino, A relative spannedness for log canonical pairs and quasi-log canonical pairs, Ann. Sc. Norm. Super. Pisa Cl. Sci. (5) **23** (2022), no. 1, 265–292.

[Fu11] O. Fujino, Cone theorem and Mori hyperbolicity, to appear in J. Differential Geom.

[Fu12] O. Fujino, On quasi-log schemes, J. Math. Soc. Japan **75** (2023), no. 3, 829–856.

[Fu13] O. Fujino, Minimal model program for projective morphisms between complex analytic spaces, preprint (2022). arXiv:2201.11315 [math.AG]

[Fu14] O. Fujino, ACC for log canonical thresholds for complex analytic spaces, to appear in *Higher Dimensional Algebraic Geometry, A Volume in Honor of V. V. Shokurov*, London Mathematical Society Lecture Note Series, **489**. Cambridge University Press, Cambridge.

[Fu15] O. Fujino, Vanishing theorems for projective morphisms between complex analytic spaces, to appear in Math. Res. Lett.

[Fu16] O. Fujino, On quasi-log structures for complex analytic spaces, preprint (2023). arXiv:2209.11401 [math.AG]

[Fu17] O. Fujino, Log canonical inversion of adjunction, Proc. Japan Acad. Ser. A Math. Sci. **100** (2024), no. 2, 7–11.

[Fu18] O. Fujino, On vanishing theorems for analytic spaces, Proc. Japan Acad. Ser. A Math. Sci. **100** (2024), no. 4, 21–25.

[Fu19] O. Fujino, On finiteness of relative log pluricanonical representations, preprint (2024).

[FF] O. Fujino, T. Fujisawa, Variation of mixed Hodge structure and its applications, preprint (2023). arXiv:2304.00672 [math.AG]

[FFS] O. Fujino, T. Fujisawa, M. Saito, Some remarks on the semipositivity theorems, Publ. Res. Inst. Math. Sci. **50** (2014), no. 1, 85–112.

[FG1] O. Fujino, Y. Gongyo, Log pluricanonical representations and the abundance conjecture, Compos. Math. 150 (2014), no. 4, 593–620.

[FG2] O. Fujino, Y. Gongyo, On log canonical rings, *Higher dimensional algebraic geometry—in honour of Professor Yujiro Kawamata's sixtieth birthday*, 159–169, Adv. Stud. Pure Math., **74**, Math. Soc. Japan, Tokyo, 2017.

[FM1] O. Fujino, K. Miyamoto, A characterization of projective spaces from the Mori theoretic viewpoint, Osaka J. Math. **58** (2021), no. 4, 827–837.

[FM2] O. Fujino, K. Miyamoto, Nakai–Moishezon ampleness criterion for real line bundles, Math. Ann. **385** (2023), no. 1-2, 459–470.

[HM] C. D. Hacon, J. McKernan, Existence of minimal models for varieties of log general type. II, J. Amer. Math. Soc. **23** (2010), no. 2, 469–490.

[Har] R. Hartshorne, *Ample subvarieties of algebraic varieties*, Notes written in collaboration with C. Musili, Lecture Notes in Mathematics, Vol. **156**. Springer-Verlag, Berlin-New York, 1970.

[Has1] K. Hashizume, Remarks on special kinds of the relative log minimal model program, Manuscripta Math. **160** (2019), no. 3-4, 285–314.

[Has2] K. Hashizume, Finiteness of log abundant log canonical pairs in log minimal model program with scaling, to appear in Michigan Math. J.

[HasH] K. Hashizume, Z. Hu, On minimal model theory for log abundant lc pairs, J. Reine Angew. Math. **767** (2020), 109–159.

[Hi] H. Hironaka, Flattening theorem in complex-analytic geometry, Amer. J. Math. **97** (1975), 503–547.

[Ka] Y. Kawamata, On the length of an extremal rational curve, Invent. Math. **105** (1991), no. 3, 609–611.

[KMM] Y. Kawamata, K. Matsuda, K. Matsuki, Introduction to the minimal model problem, *Algebraic geometry, Sendai, 1985*, 283–360, Adv. Stud. Pure Math., **10**, North-Holland, Amsterdam, 1987.

[Kod] K. Kodaira, *Complex manifolds and deformation of complex structures*, Translated from the 1981 Japanese original by Kazuo Akao. Reprint of the 1986 English edition. Classics in Mathematics. Springer-Verlag, Berlin, 2005.

[Kol] J. Kollár, *Rational curves on algebraic varieties*, Ergebnisse der Mathematik und ihrer Grenzgebiete. 3. Folge. A Series of Modern Surveys in Mathematics [Results in Mathematics and Related Areas. 3rd Series. A Series of Modern Surveys in Mathematics], **32**. Springer-Verlag, Berlin, 1996.

[KM] J. Kollár, S. Mori, *Birational geometry of algebraic varieties*, With the collaboration of C. H. Clemens and A. Corti. Translated from the 1998 Japanese original. Cambridge Tracts in Mathematics, **134**. Cambridge University Press, Cambridge, 1998.

[Le] J. Lesieutre, Notions of numerical Iitaka dimension do not coincide, J. Algebraic Geom. **31** (2022), no. 1, 113–126.

[LM] S. Lyu, T. Murayama, The relative minimal model program for excellent algebraic spaces and analytic spaces in equal characteristic zero, preprint (2022). arXiv:2209.08732 [math.AG]

[Ma] K. Matsuki, *Introduction to the Mori program*, Universitext. Springer-Verlag, New York, 2002.

[Mo] S. Mori, Threefolds whose canonical bundles are not numerically effective, Ann. of Math. (2) **116** (1982), no. 1, 133–176.

[Mu] T. Murayama, Injectivity theorems and cubical descent for schemes, stacks, and analytic spaces, preprint (2024). arXiv:2406.10800 [math.AG]

[Na1] N. Nakayama, The lower semicontinuity of the plurigenera of complex varieties, *Algebraic geometry, Sendai, 1985*, 551–590, Adv. Stud. Pure Math., **10**, North-Holland, Amsterdam, 1987.

[Na2] N. Nakayama, *Zariski-decomposition and abundance*, MSJ Memoirs, **14**. Mathematical Society of Japan, Tokyo, 2004.

[OSS] C. Okonek, M. Schneider, H. Spindler, *Vector bundles on complex projective spaces*, Corrected reprint of the 1988 edition. With an appendix by S. I. Gelfand. Modern Birkhäuser Classics. Birkhäuser/Springer Basel AG, Basel, 2011.

[Sa1] M. Saito, Modules de Hodge polarisables, Publ. Res. Inst. Math. Sci. **24** (1988), no. 6, 849–995.

[Sa2] M. Saito, Mixed Hodge modules, Publ. Res. Inst. Math. Sci. **26** (1990), no. 2, 221–333.

[Sa3] M. Saito, Decomposition theorem for proper Kähler morphisms, Tohoku Math. J. (2) **42** (1990), no. 2, 127–147.

[Sa4] M. Saito, Some remarks on decomposition theorem for proper Kähler morphisms, preprint (2022). arXiv:2204.09026 [math.AG]

[Se] J.-P. Serre, Géométrie algébrique et géométrie analytique, Ann. Inst. Fourier (Grenoble) **6** (1955/56), 1–42.

[Sh] I. R. Shafarevich, *Basic algebraic geometry. 2. Schemes and complex manifolds*, Third edition. Translated from the 2007 third Russian edition by Miles Reid. Springer, Heidelberg, 2013.

[Si] Y.-T. Siu, Noether–Lasker decomposition of coherent analytic subsheaves, Trans. Amer. Math. Soc. **135** (1969), 375–385.

[Ta] K. Takegoshi, Higher direct images of canonical sheaves tensorized with semi-positive vector bundles by proper Kähler morphisms, Math. Ann. **303** (1995), no. 3, 389–416.

Index